NATIONAL ACADEMIES
Sciences
Engineering
Medicine

NATIONAL ACADEMIES PRESS
Washington, DC

Analysis of Potential Interference Issues Related to FCC Order 20-48

Committee to Review FCC Order 20-48 Authorizing Operation of a Terrestrial Radio Network Near the GPS Frequency Bands

Computer Science and Telecommunications Board

Air Force Studies Board

Board on Physics and Astronomy

Division on Engineering and Physical Sciences

NATIONAL ACADEMIES PRESS 500 Fifth Street, NW Washington, DC 20001

This activity was supported by the U.S. Department of Defense Chief Information Officer with the assistance of the Air Force Research Laboratory under award number FA955016D0001/FA8650-21-F-9309. Any opinions, findings, conclusions, or recommendations expressed in this publication do not necessarily reflect the views of any organization or agency that provided support for the project.

International Standard Book Number-13: 978-0-309-69007-2
International Standard Book Number-10: 0-309-69007-2
Digital Object Identifier: https://doi.org/10.17226/26611

This publication is available from the National Academies Press, 500 Fifth Street, NW, Keck 360, Washington, DC 20001; (800) 624-6242 or (202) 334-3313; http://www.nap.edu.

Copyright 2023 by the National Academy of Sciences. National Academies of Sciences, Engineering, and Medicine and National Academies Press and the graphical logos for each are all trademarks of the National Academy of Sciences. All rights reserved.

Printed in the United States of America.

Suggested citation: National Academies of Sciences, Engineering, and Medicine. 2023. *Analysis of Potential Interference Issues Related to FCC Order 20-48*. Washington, DC: The National Academies Press. https://doi.org/10.17226/26611.

The **National Academy of Sciences** was established in 1863 by an Act of Congress, signed by President Lincoln, as a private, nongovernmental institution to advise the nation on issues related to science and technology. Members are elected by their peers for outstanding contributions to research. Dr. Marcia McNutt is president.

The **National Academy of Engineering** was established in 1964 under the charter of the National Academy of Sciences to bring the practices of engineering to advising the nation. Members are elected by their peers for extraordinary contributions to engineering. Dr. John L. Anderson is president.

The **National Academy of Medicine** (formerly the Institute of Medicine) was established in 1970 under the charter of the National Academy of Sciences to advise the nation on medical and health issues. Members are elected by their peers for distinguished contributions to medicine and health. Dr. Victor J. Dzau is president.

The three Academies work together as the **National Academies of Sciences, Engineering, and Medicine** to provide independent, objective analysis and advice to the nation and conduct other activities to solve complex problems and inform public policy decisions. The National Academies also encourage education and research, recognize outstanding contributions to knowledge, and increase public understanding in matters of science, engineering, and medicine.

Learn more about the National Academies of Sciences, Engineering, and Medicine at **www.nationalacademies.org**.

Consensus Study Reports published by the National Academies of Sciences, Engineering, and Medicine document the evidence-based consensus on the study's statement of task by an authoring committee of experts. Reports typically include findings, conclusions, and recommendations based on information gathered by the committee and the committee's deliberations. Each report has been subjected to a rigorous and independent peer-review process and it represents the position of the National Academies on the statement of task.

Proceedings published by the National Academies of Sciences, Engineering, and Medicine chronicle the presentations and discussions at a workshop, symposium, or other event convened by the National Academies. The statements and opinions contained in proceedings are those of the participants and are not endorsed by other participants, the planning committee, or the National Academies.

Rapid Expert Consultations published by the National Academies of Sciences, Engineering, and Medicine are authored by subject-matter experts on narrowly focused topics that can be supported by a body of evidence. The discussions contained in rapid expert consultations are considered those of the authors and do not contain policy recommendations. Rapid expert consultations are reviewed by the institution before release.

For information about other products and activities of the National Academies, please visit www.nationalacademies.org/about/whatwedo.

Committee to Review FCC Order 20-48 Authorizing Operation of a Terrestrial Radio Network Near the GPS Frequency Bands

J. MICHAEL McQUADE, Carnegie Mellon University, *Chair*
JENNIFER L. ALVAREZ, Aurora Insight, Inc.
KRISTINE M. LARSON (NAS), University of Colorado Boulder
JOHN L. MANFERDELLI, VMware
PRESTON F. MARSHALL, Google, LLC
MARK L. PSIAKI, Virginia Polytechnic Institute and State University
RICHARD L. REASER, JR., Independent Consultant
JEFFREY H. REED, Virginia Polytechnic Institute and State University
NAMBIRAJAN SESHADRI (NAE), University of California, San Diego
J. SCOTT STADLER, Massachusetts Institute of Technology Lincoln Laboratory
STEPHEN J. STAFFORD, Johns Hopkins University Applied Physics Laboratory

Staff

JON EISENBERG, Senior Director, Computer Science and Telecommunications Board
RYAN MURPHY, Program Officer, Air Force Studies Board
GREGORY MACK, Senior Program Officer, Board on Physics and Astronomy (through January 2022)
SHENAE BRADLEY, Administrative Assistant, Computer Science and Telecommunications Board
KATIRIA ORTIZ, Associate Program Officer, Computer Science and Telecommunications Board

COMPUTER SCIENCE AND TELECOMMUNICATIONS BOARD

LAURA HAAS (NAE), University of Massachusetts Amherst, *Chair*
DAVID CULLER (NAE), University of California, Berkeley
ERIC HORVITZ (NAE), Microsoft Research
CHARLES ISBELL, Georgia Institute of Technology
ELIZABETH MYNATT, Georgia Institute of Technology
CRAIG PARTRIDGE, Colorado State University
DANIELA RUS (NAE), Massachusetts Institute of Technology
MARGO SELTZER (NAE), University of British Columbia
NAMBIRAJAN SESHADRI (NAE), University of California, San Diego
MOSHE Y. VARDI (NAS/NAE), Rice University

Staff

JON K. EISENBERG, Senior Board Director
SHENAE A. BRADLEY, Administrative Assistant
RENEE HAWKINS, Finance Business Partner
THƠ NGUYỄN, Senior Program Officer
KATIRIA ORTIZ, Associate Program Officer
BRENDAN ROACH, Program Officer

Preface

Section 1663 of the Fiscal Year (FY) 2021 National Defense Authorization Act called on the U.S. Department of Defense (DoD) to enter into an agreement with the National Academies of Sciences, Engineering, and Medicine to carry out "an independent technical review of the order and authorization adopted by the Federal Communications Commission on April 19, 2020 (FCC 20-48), to the extent that such Order and Authorization affects the devices, operations, or activities of the Department of Defense." The Office of the Secretary of Defense, Chief Information Officer, with the assistance of the Air Force Research Laboratory, entered into a contract with the National Academies, and the National Academies appointed the Committee to Review FCC Order 20-48 Authorizing Operation of a Terrestrial Radio Network Near the GPS Frequency Bands (committee member biographical information can be found in Appendix D) to carry out the study per the statement of task in Appendix A.

The committee met on a nearly weekly basis from September 2021 to April 2022 to plan the study; receive briefings from experts and stakeholders (Appendix B); and review relevant reports, technical literature, and written submissions to the committee (Appendix C). Materials submitted to the committee can be obtained from the National Academies' Public Access Records Office.[1] In addition, a cleared subset of the committee received a set of classified briefings that informed a classified annex to this report.

---

[1] Complete the National Academies of Sciences, Engineering, and Medicine "Request for Information from the Public Access Records Office" form for this project at https://www8.nationalacademies.org/pa/managerequest.aspx?key=DEPS-CSTB-21-02.

Reviewers

This Consensus Study Report was reviewed in draft form by individuals chosen for their diverse perspectives and technical expertise. The purpose of this independent review is to provide candid and critical comments that will assist the National Academies of Sciences, Engineering, and Medicine in making each published report as sound as possible and to ensure that it meets the institutional standards for quality, objectivity, evidence, and responsiveness to the study charge. The review comments and draft manuscript remain confidential to protect the integrity of the deliberative process.

We thank the following individuals for their review of this report:

ALISON K. BROWN (NAE), NAVSYS Corporation
ALDEN FUCHS, Fuchs Consulting
CHRISTOPHER HEGARTY, The MITRE Corporation
TODD HUMPHREYS, The University of Texas at Austin
MARK LOFQUIST, The Aerospace Corporation and University of
    Colorado Boulder
DOUGLAS LOVERRO, Loverro Consulting
JON PEHA, Carnegie Mellon University
WILLIAM PRESS (NAS), The University of Texas at Austin
DOROTHY ROBYN, Boston University
DOUGLAS SICKER, University of Colorado Denver

Although the reviewers listed above provided many constructive comments and suggestions, they were not asked to endorse the conclusions or recommendations of this report nor did they see the final draft before its release. The review of this report was overseen by STEVEN J. BATTEL (NAE), Battel Engineering, and AROGYASWAMI J. PAULRAJ (NAE), Stanford University. They were responsible for making certain that an independent examination of this report was carried out in accordance with the standards of the National Academies and that all review comments were carefully considered. Responsibility for the final content rests entirely with the authoring committee and the National Academies.

Contents

SUMMARY   1

1   INTRODUCTION AND BACKGROUND   9
    1.1   Spectrum Overview, 9
    1.2   GPS Overview, 13
    1.3   FCC Order 20-48, 27
    1.4   Harmful Interference, 32

2   ANALYSIS REGARDING THE THREE STUDY TASKS   50
    2.1   Task 1: Approaches to Evaluating Harmful Interference Concerns, 50
    2.2   Task 2: Harmful Interference to GPS and Mobile Satellite Services, 57
    2.3   Task 3: Feasibility, Practicality, and Effectiveness of Mitigation Measures in the FCC Order, 74

3   ADDITIONAL MATTERS   78
    3.1   Toward a Better Means of Assessing Harmful Interference, 78
    3.2   Ways to Manage Potential Future Controversies, 81

4   CONCLUDING THOUGHTS   86

APPENDIXES

A   STATEMENT OF TASK   91
B   PRESENTATIONS TO THE COMMITTEE   93
C   ORGANIZATIONS AND INDIVIDUALS THAT PROVIDED WRITTEN INPUT TO THE COMMITTEE   96
D   COMMITTEE MEMBER BIOGRAPHICAL INFORMATION   97
E   DISCLOSURE OF UNAVOIDABLE CONFLICTS OF INTEREST   103

Summary

BACKGROUND

On April 19, 2020, the Federal Communications Commission (FCC) adopted order 20-48 authorizing Ligado Networks, LLC, to move forward with the deployment and operation of a low-power terrestrial nationwide radio network. The order authorizes Ligado to provide terrestrial base station to mobile earth station downlink transmissions in the 1526–1536 MHz band within the mobile satellite services (MSS)-allocated 1525–1559 MHz band and mobile earth station to terrestrial base station uplink transmissions in the 1627.5–1637.5 and 1646.5–1656.5 MHz bands of the 1626.5–1660.5 MHz band. The order comes after a process that began 17 years earlier when Ligado's predecessor-in-interest (AMSC Subsidiary Corporation) originally obtained certain MSS licenses. This process has involved multiple application modifications and the commissioning of numerous test-and-evaluation and analytical studies. The complexity of this region of the spectrum, including its adjacency to Global Positioning System (GPS) and other satellite services and the variety of commercial, economic, and national security interests involved has led some parties to question the analyses and assumptions underlying the FCC order. To help shed light on the dispute, the U.S. Congress requested a study by the National Academies of Sciences, Engineering, and Medicine.

The motivation for this study and report is the possibility that emissions by the Ligado system as it would be deployed in accordance with the terms authorized in FCC Order 20-48 might disrupt GPS services or MSS. At issue are two simple facts: (1) radio-frequency transmitters do not operate with arbitrarily sharp cutoff frequencies and thus, depending on how rapidly (as a function of frequency) their emitted power spectrum falls off, may emit power beyond their authorized spectral bands; and (2) receivers of electromagnetic spectrum do not "listen" only within a band defined with arbitrarily sharp boundaries and thus may receive power from frequencies outside their designed band. The FCC's goal is to enable as much productive use of valuable spectrum resources as possible, balanced against the deleterious impacts that may arise when adjacent signals cause receivers to experience interference.

GPS services are vital to the modern economy and to national defense operations. Standard receivers provide positioning information at the meter scale and precision timing at the tens-of-nanosecond scale. High-precision receivers, such as those used in surveying, geodesy, and infrastructure and farming applications, can deliver positioning services at the centimeter or smaller scale. Interference of the GPS receivers can potentially lead to degraded performance or loss of operation.

This study was not charged with considering whether the FCC reached a correct outcome in authorizing the Ligado system. That is the purview of the FCC's processes involving materials and testimony from a wide array of interested parties over the course of the proceedings. Instead, the committee was asked to consider three specific tasks, the response to which make up the bulk of this study. The three principal tasks for the committee, paraphrasing the statement of task (see Appendix A), are as follows:

- **Task 1:** Assess which of two commonly used approaches to evaluating interference that might cause harm to GPS services would most effectively mitigate the risks of harmful interference to GPS services and U.S. Department of Defense (DoD) operations and activities.
- **Task 2:** Assess the likelihood that the authorized Ligado service will create harmful interference to GPS, MSS, and other commercial or DoD services and operations.
- **Task 3:** Assess the feasibility, practicality, and effectiveness of the measures in the FCC order to mitigate harmful interference effects on DoD devices, operations, and activities.

In providing responses to these tasks, the committee stresses three important points:

- Throughout this report, the term "Harmful Interference"—when capitalized—is a defined term used by the FCC and is specifically not the same thing as the general term "interference," which describes what happens in a receiver when some other signal affects the intended received signal in a way that reduces the effective received signal-to-noise ratio (SNR). This report uses the uncapitalized term "harmful interference" in a more general sense to imply degraded receiver operations without assessing whether such degradation is actually causing a degradation of function or whether the receiver is operating in accordance with FCC rules. The committee assumed that in the statement of task, "harmful interference" meant the defined term.

- The legal definition of "Harmful Interference" also includes whether affected devices are or are not "operating in accordance with [the International Telecommunication Union] Radio Regulations" and "operating in accordance with this chapter." However, the committee proceeded under the assumption that it has been charged with determining whether existing radionavigation satellite services (RNSS) or MSS would be harmed by Ligado interference independent of any legal ruling and concentrated on the physics and engineering questions of harmful interference.
- FCC Order 20-48 authorizes Ligado's use of spectrum in the United States only. The committee did not assess and reached no conclusion regarding the potential for interference caused by emissions in the bands in question outside the United States.

ANSWERS TO THE THREE ELEMENTS OF THE STUDY TASK

Task 1: Approaches to Evaluating Harmful Interference Concerns

Task 1 asks which of the two prevailing proposed approaches to evaluating harmful interference concerns—one based on a signal-to-noise interference protection criterion (IPC) and the other based on a device-by-device measurement of the GPS position error—most effectively mitigates risks of harmful interference with GPS services and DoD operations and activities.

Conclusion 1: Neither of the prevailing approaches to evaluating harmful interference concerns effectively mitigates the risk of harmful interference.

Neither approach provides analytical, repeatable, or straightforward criteria to evaluate new entrants. Both approaches have a role to play in evaluating harmful interference to existing receivers. The signal-to-noise approach is inflexible as a determinant or threshold, providing what in some circumstances may be an overly conservative emission limit because no single value for signal-to-noise degradation determines when the various types of possible harm to receiver performance will become significant. The position measurement approach is dependent on the test sampling approach and is too narrow in its applicability to the many and varied uses of the GPS.

Of the two approaches, the one based on the SNR, when done properly, is the more comprehensive and informative. By indicating the degradation in link margin, this approach can be used to predict harmful impact across a broad set of use cases. However, the commonly advocated 1 dB SNR loss criterion has not been linked to the FCC's definition of Harmful Interference. Although adherence to a "1 dB criterion" may generally prevent Harmful Interference, the vast majority of GPS use cases do not experience Harmful Interference at that level. As such, the 1 dB criterion is prophylactic, but conservative.

The determination of Harmful Interference is dependent on the characteristics of the transmitter and the receiver as well as the particulars of each specific use. There is a wide array of GPS use cases—for example, car navigation, network timing, precise surveying and farming, geophysical monitoring, and aircraft ground and flight operations—all of which may be relevant to DoD operations. These use cases have different failure modes, which result in varying interference tolerance. The most appropriate approach to evaluating potentially harmful interference must be mapped to each relevant use case. For example, some applications are harmed when code-lock is lost, while other applications are harmed from loss of carrier phase lock. As such, the use of a single, fixed SNR-based IPC is not practical when applied to device-by-device performance.

The question posed in the task does not directly address the bigger challenge: regardless of which approach is applied, drastically different conclusions can be reached. There are numerous test design particulars that must be considered, including determining which path loss model to use, the appropriate stand-off range to use, antenna coupling, the degree of insensitivity of a particular receiver's design to adjacent-band power, and performance thresholds. Even for a given use case, these issues are not easily resolved. Furthermore, a per-device SNR threshold may create a downstream hazard—receiver manufacturers may be incentivized to keep using outdated designs that preclude higher-value use of nearby spectrum or to design adjacent-band-susceptible receivers to claim spectral easements.

Ultimately, both proposed approaches are cumbersome, owing to the intensive, device-by-device testing required. They do not provide an engineerable, predictable standard that new entrants can readily use to evaluate impact. As such, these approaches impede progress in making more efficient and effective use of the spectrum. A new applicant for emissions in a new adjacent channel will have great difficulty in determining the emitter power levels and stand-off distances that will be guaranteed not to cause Harmful Interference to the installed base of GPS receivers. A GPS receiver designer will be unable to design a receiver that will be guaranteed to tolerate unknown potential future allowed levels of adjacent-band power.

Task 2: Harmful Interference to GPS and Mobile Satellite Services

Task 2 asks about the potential for harmful interference from the proposed Ligado network to GPS, MSS, and other commercial or DoD services, including the potential to affect DoD operations and activities. This potential is evaluated across several different use cases, each restricted to operation in the United States.

For GPS, several sets of interference tests have been performed that span many representative GPS devices drawn from many different receiver classes and suppliers. The tests evaluated various scenarios and advocated for different metrics to determine the onset of harmful interference. Despite these differences, the results consistently indicate that a majority of devices do not experience harmful interference. This is discussed in more detail in Section 2.2.

> *Conclusion 2: Based on the results of tests conducted to inform the Ligado proceeding, most commercially produced general navigation, timing, cellular, or certified aviation GPS receivers will not experience significant harmful interference from Ligado emissions as authorized by the FCC. High-precision receivers are the most vulnerable receiver class, with the largest proportion of units tested that will experience significant harmful interference from Ligado operations as authorized by the FCC.*

The committee also reached the following conclusion regarding the state of the art in GPS receivers:

> *Conclusion 3: It is within the state-of-the-practice of current technology to build a receiver that is robust to Ligado signals for any GPS application, and all GPS receiver manufacturers could field new designs that could co-exist with the authorized Ligado signals and achieve good performance even if their existing designs cannot.*

Turning to impacts on MSS, the committee concluded that the Globalstar system is unlikely to experience harmful interference, because only its uplink is in the L-band and it uses code-division multiple access (CDMA) signals. However:

> *Conclusion 4: Iridium terminals will experience harmful interference on their downlink caused by Ligado user terminals operating in the UL1 band while those Iridium terminals are within a significant range of a Ligado emitter—up to 732 meters.*

Additionally, DoD has evaluated the impact of FCC Order 20-48 on department devices and missions. The following summary points were provided to the committee in a set of slides dated March 15, 2022.[1] It is important to note that the following conclusions were asserted by DoD without providing publicly available supporting data and were not discussed by the committee in a public session:

- DoD and interagency partners conducted testing to determine the impacts to GPS (captures FCC Order 20-48's authorized deployment). The tests demonstrated that the proposed signal introduces harmful interference to critical national security mission capabilities.
- The terrestrial network authorized by FCC Order 20-48 will create unacceptable harmful interference for DoD missions. The mitigation techniques and other regulatory provisions in FCC Order 20-48 are insufficient to protect national security missions.

Additional information on the test results and analysis as they related to DoD systems and missions is discussed in a classified annex to this report.

Task 3: Feasibility, Practicality, and Effectiveness of Mitigation Measures in the FCC Order

Task 3 asks about the feasibility, practicality, and effectiveness of the mitigation measures required in the FCC order with respect to DoD devices, operations, and activities. The FCC order enumerates several potential mitigations when a receiver experiences harmful interference, including but not limited to enacting exclusion zones for Ligado emitters; replacing components (e.g., antennas or filters) or full receivers; enabling a "kill switch" mechanism for Ligado to turn off emitters in some geographic locations; and additional negotiated mitigations between Ligado and the affected government agency for Ligado to reduce emissions to an acceptable received power level over certain installations.

For the question of harmful interference with GPSs, the effectiveness and practicality of any of the foregoing potential mitigations depends on the type of receiver and the application. One must also distinguish between two types of equipment:

1. *DoD-authorized/compliant devices* approved for weapons and weapons delivery systems and other national security certified devices.
2. *Commercial GPS devices* when used in national security applications with an express waiver or navigational warfare (NAVWAR) compliance determination

---

[1] These slides, along with other materials provided to the committee, were placed in the project's public access file and are available on request from the National Academies' Public Access Records Office.

per DoD Instruction 4650.08 and CJCSI 6140-01 or commercial devices that are used in other DoD operations or missions such as emergency response or partner operations.

For DoD authorized/compliant devices, GPS receivers, or systems that incorporate such GPS receivers used inside the United States, replacing older devices with newer versions of such devices that are protected from harmful interference and already qualified may provide a plausible solution. However, where such replacements are not immediately available this study concludes that this is not likely a satisfactory mitigation. These systems typically must pass very long and expensive operational test certification; generally, mitigations that include replacing or augmenting older devices would involve unsatisfactorily long delays.

Regarding commercial devices used by DoD, the committee concludes that several of the mitigation procedures may be effective but only to the extent that they are timely, affordable, and practicable. (The question of affordability may, in the end, be a question for Ligado and not DoD because Ligado has asserted that they have committed to bear the cost of mitigating such DoD interference problems.) Such actions are highly application specific. These include replacing antenna subsystems; full-scale replacement of older commercial GPS receivers with newer models; and negotiated extended exclusion zones in which no Ligado emitters are placed.

Conclusion 5: Although the mitigation procedures proposed in the order may be effective, in many cases such mitigation may be impractical without the extensive dialog among the affected parties presumed in the order. In some cases, mitigation may not be practical at operationally relevant time scales or at reasonable cost.

FCC Order 20-48 sets forth procedures for Ligado and those claiming interference to engage in dialog to determine if Ligado is, in fact, responsible for causing harmful interference to a DoD-operated receiver. As noted in this report's analysis of inference and in commentary on Tasks 1 and 2 (see Sections 1.4, 2.1, and 2.2), interference is not simple, and receivers live in complicated electromagnetic environments. As a result, mitigation will not be practical without extensive dialog as intended within the order. Even if such dialog takes place, mitigation may in some cases not be practicable within operationally relevant time and financial parameters.

ADDITIONAL OBSERVATIONS

The study committee's statement of task (see Appendix A) also provides that the committee may address "other related issues the study committee determines relevant." In the course of its work, the committee has concluded that there are several important issues surrounding the technical and administrative processes used in the long saga that has led to the authorities granted in FCC Order 20-48. Chapter 3 provides observations with the hope that future proceedings might provide more streamlined and optimized approaches to balance protection of incumbents with maximizing the economic and operational benefits provided by new entrants into a given spectrum region.

Spectrum real estate is a living asset, and approaches must allow not only for a degree of confidence that a deployed system will not be compromised by future, unforeseen entrants, *for a period of time*, but also must recognize that capabilities will evolve. Some form of more definitive receiver standards and establishment of set time periods where adherence to those receiver standards will ensure successful operation for a frequency band's incumbents and new entrants seem to be important tools in this regard.

1

Introduction and Background

This chapter provides several background sections necessary to the understanding of the committee's responses to its three tasks. The chapter begins with an overview of the region of the electromagnetic spectrum relevant to this study (Section 1.1); followed by a brief tutorial on the Global Positioning System, or GPS (Section 1.2); and then a summary of the specific authorizations and mitigation procedures in Federal Communications Commission (FCC) Order 20-48 (Section 1.3). Chapter 1 closes with a discussion of what is meant by interference, why it is a physics phenomenon, and why it is not the same thing as the defined term "Harmful Interference" (Section 1.4).

1.1 SPECTRUM OVERVIEW

The radio frequency spectrum is a natural resource that underpins all wireless activity, and as a critical resource, is managed to ensure its effective use. In the United States, the National Telecommunications and Information Administration (NTIA) establishes policy, manages spectrum, and assigns frequencies for federal government users, and the FCC regulates spectrum for all other users.

1.1.1 Allocations Under Consideration

Spectrum is divided into frequency bands that are allocated for specific radio services. The allocations under consideration in FCC Order 20-48 primarily pertain to two radio services in the frequency range from 1525 MHz to 1660.5 MHz: mobile satellite services (MSS) and radionavigation satellite services (RNSS).

MSS is defined by the International Telecommunication Union (ITU) as "a radio-communication service between mobile Earth stations and one or more space stations, or between space stations used by this service; or between mobile Earth stations by means of one or more space stations."[1]

RNSS is defined as "a radiodetermination-satellite service used for the purpose of radionavigation."[2] Radiodetermination is defined as "the determination of the position, velocity and/or other characteristics of an object, or the obtaining of information relating to these parameters, by means of the propagation properties of radio waves."[3]

1.1.2 Frequency Bands Under Consideration

MSS space-to-Earth downlinks are from 1525 to 1559 MHz, and MSS Earth-to-space uplinks are from 1610 to 1660.5 MHz, with a downlink allocation also from 1613.8 to 1626.5 MHz. The L5 band is centered at 1176.45 MHz. RNSS is from 1559 to 1610 MHz.[4] Figure 1-1 depicts these allocations.

Within the MSS allocations, licenses are assigned to particular users. MSS licensed frequency bands under consideration are shown in Figure 1-2 and include

- Ligado:
    - 1526–1536 MHz (downlink)
    - 1627.5–1637.5 MHz (uplink)
    - 1646.5–1656.5 MHz (uplink)[5]
    - 1670–1680 MHz (proposed downlink)

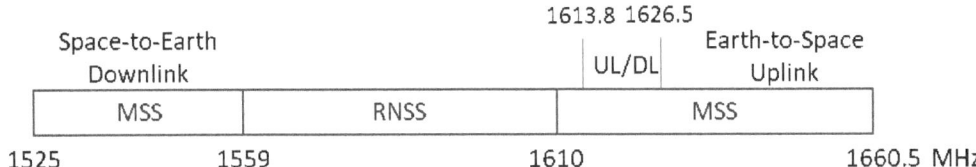

FIGURE 1-1 Allocations for mobile satellite services (MSS) and radionavigation satellite services (RNSS) from 1525 to 1660.5 MHz.
NOTE: DL, downlink; UL, uplink.

---

[1] International Telecommunication Union, 2020, "Radio Regulations Articles Edition of 2020," item 1.25.
[2] International Telecommunication Union, 2020, "Radio Regulations Articles Edition of 2020," item 1.43.
[3] International Telecommunication Union, 2020, "Radio Regulations Articles Edition of 2020," item 1.9.
[4] National Telecommunications and Information Administration, 2017, *Manual of Regulations and Procedures for Federal Radio Frequency Management*, Washington, DC: U.S. Department of Commerce, September.
[5] Federal Communications Commission, 2020, FCC Order 20-48, Order and Authorization, April 22; hereafter, FCC Order 20-48.

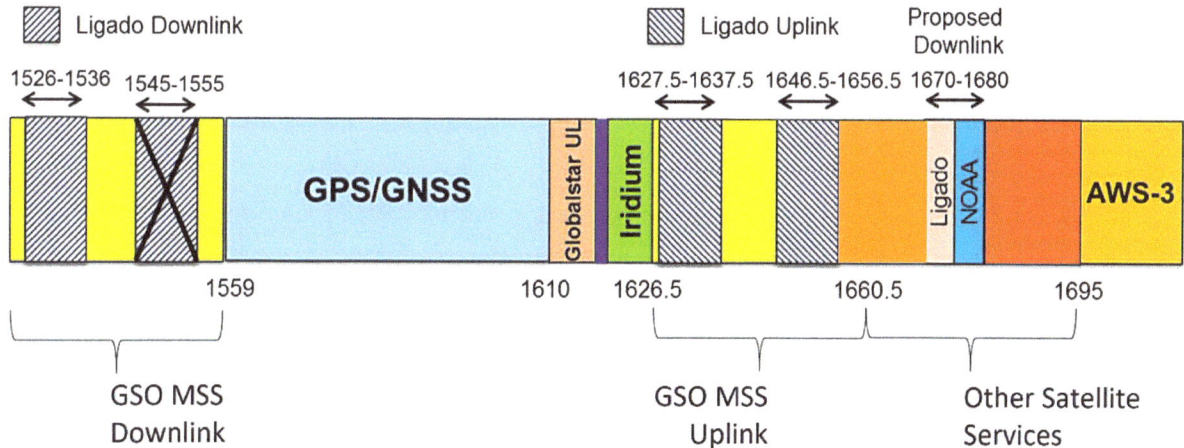

FIGURE 1-2 The Global Positioning System (GPS) adjacent spectrum, showing the various frequency allocations. Note that an earlier version of the Ligado proposal had a downlink using the 1545–1555 MHz band, but that was not authorized in Federal Communications Commission (FCC) Order 20-48.
NOTE: AWS, Advanced Wireless Services; GNSS, Global Navigation Satellite System; GSO, geostationary orbit; MSS, mobile satellite services; NOAA, National Oceanic and Atmospheric Administration; UL, uplink.
SOURCE: Coalition of Aviation, SATCOM, and Weather Information Users, "Impact of Ligado's Proposal on SATCOM, Aviation and Weather Data Users," Notice of Exparte, FCC Proceedings IB 12-340 and IB 11-109, Coalition Deck for Sept. 4 2019 FCC meeting, posted September 9, 2019, https://www.fcc.gov/ecfs/document/10906015584180/3.

- Iridium: 1617.775–1626.5 MHz (uplink and downlink)[6]
- Globalstar: 1610–1618.725 MHz (uplink)
- Inmarsat: 1626.5–1660.5 MHz (uplink)

The GPS band specifically under consideration is the L1 signal centered at 1575.42 MHz with a bandwidth of 30.69 MHz (1560.075–1590.765 MHz).[7]

An additional element in the general description of MSS is the introduction of a service called Ancillary Terrestrial Component (ATC), which permits MSS licensees to operate, if approved and licensed, terrestrial base stations and mobile terminals for the purpose of filling gaps in service areas.[8] Ligado was granted authorization for ATC operations in its MSS-licensed bands.[9]

---

[6] M. McLaughlin, 2021, "Iridium Communications Inc. Letter Regarding NAS Committee Members [DEPSCSTB-21-02]—Review of FCC Order 20-48 Authorizing Operation of a Terrestrial Radio Network Near the GPS Frequency Bands," letter submitted to the Committee to Review FCC Order 20-48, Washington, DC: National Academies of Sciences, Engineering, and Medicine.

[7] Interface Control Working Group, 2021, "NAVSTAR GPS Space Segment/Navigation User Interfaces," IS-GPS-200 Revision M, April 13, approved May 21, El Segundo, CA: SAIC.

[8] National Telecommunications and Information Administration, 2021, *Manual of Regulations and Procedures for Federal Radio Frequency Management*, Washington, DC: U.S. Department of Commerce, Chapter 4c, pp. 4–236. US380 In the bands 1525–1544 MHz, 1545–1559 MHz, 1610–1645.5 MHz, 1646.5–1660.5 MHz, and 2483.5–2500 MHz, a non-federal licensee in the mobile satellite services (MSS) may also operate an ancillary terrestrial component in conjunction with its MSS network, subject to the Commission's rules for ancillary terrestrial components and subject to all applicable conditions and provisions of its MSS authorization.

[9] See FCC Order 20-48. It should be noted that the order authorizes a nationwide network, not merely "filling in the gaps."

1.1.3 In-Band and Out-of-Band Concepts

Of particular importance to the efforts under FCC Order 20-48 are the concepts of in-band emissions, in-band reception, out-of-band emissions, and out-of-band reception:

- *In-band emissions*—emissions at a frequency or frequencies for which an emission is licensed or otherwise authorized.
- *In-band reception*—emissions received at a frequency or frequencies for which a system is licensed or otherwise authorized.
- *Out-of-band emissions (OOBE)*—emission at a frequency or frequencies immediately outside the licensed band, but excluding spurious emissions.[10]
- *Out-of-band reception (OOBR)*—emissions received from beyond the licensed or otherwise authorized frequencies of a system.

These concepts are depicted in Figure 1-3.

In-band emissions are radiated emissions from devices that are licensed to transmit in the particular band of interest. For example, a Ligado base station transmitting in its MSS/ATC licensed downlink band of 1526–1536 MHz is creating an in-band emission.

Emissions, however, may not necessarily be confined to the band of interest. Owing to equipment design and limitations of active and passive electronic components within the equipment, in practice, equipment may radiate emissions beyond its designated band. These emissions are referred to as OOBE. These emissions are typically unintended and unwanted, and thus are often regulated to not cause interference to devices beyond their in-band frequency range.

Similarly, for equipment receiving a signal, equipment design and limitations of the electronics may allow emissions that are beyond the assigned in-band frequency range to be received by a device. An authorized in-band emission in an adjacent or nearby band may be received by devices that are receiving beyond their licensed band. This is termed OOBR and is generally undesirable depending on the adjacent or nearby band activity.

1.1.4 In-Band and Out-of-Band Considerations

Specific to the efforts under FCC Order 20-48 are emissions from Ligado's equipment in its licensed bands (in-band emissions) and beyond its licensed bands (OOBE). Regulatory limits on power levels have been established for Ligado emissions both in its licensed bands and beyond its licensed bands. Considerations under the order include the effects

---

[10] International Telecommunication Union, 2020, "Radio Regulations Articles Edition of 2020," item 1.144.

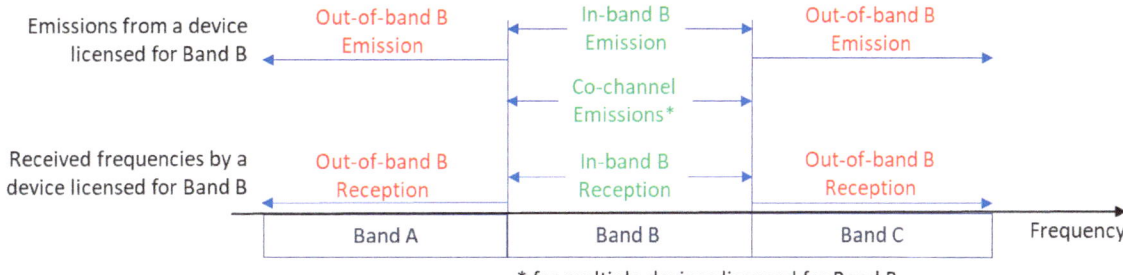

FIGURE 1-3 Concepts of in-band emissions, out-of-band emissions, in-band reception, and out-of-band reception.

of Ligado's compliant in-band emissions and compliant OOBE on users in the adjacent or nearby GPS band and MSS bands. If Ligado's equipment complies with its regulatory limits, the fundamental question is how are other users that are nearby in frequency affected?

1.2 GPS OVERVIEW

1.2.1 Background—Global Positioning System

GPS services are critical to modern society, providing coarse (5-m resolution) positioning information to billions of users, including emergency services, aviation, and travel route navigation. In addition, GPS provides attitude and precise timing for some users. More sophisticated use of GPS signals allows commercial services and scientists to locate objects with better than centimeter precision using carrier phase measurements.[11] Last, GPS is a critical technology employed in the defense of the United States and its allies.

GPS services consist of a space segment of approximately 32 satellites (space vehicles, SVs) in medium Earth orbits (~20,000 km altitude), so there are about 8 satellites "in view" at almost all locations on Earth at any given time. Each SV has a unique identifier or space vehicle number (SVN).

GPS signals are transmitted in three bands:

- L1 is centered on 1575.42 MHz, and transmitted signals are contained within a bandwidth of about 30.69 MHz.[12] L1 carries both civilian (L1 C/A) and military signals (L1 P(Y) and L1 M code), as depicted in Figure 1-4.

---

[11] P. Misra and P. Enge, 2011, *Global Positioning System: Signals, Measurements, and Performance*, Lincoln, MA: Ganga-Jamuna Press.

[12] See the GPS Interface Specification, IS-GPS-200M (Interface Control Working Group, 2021, "NAVSTAR GPS Space Segment/Navigation User Interfaces," IS-GPS-200 Revision M, April 13, approved May 21, El Segundo, CA: SAIC, https://www.gps.gov/technical/icwg/IS-GPS-200M.pdf).

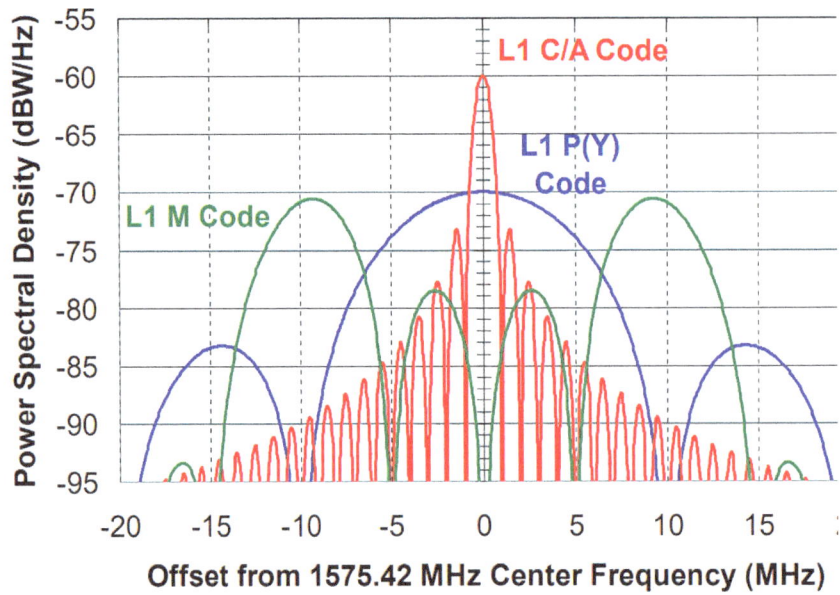

FIGURE 1-4 L1 Global Positioning Service (GPS) signal structure.
SOURCE: Underlying image from U.S. Geological Survey, 2004, "Improving the GPS L1 Signal," presentation to the National Space-Based Positioning, Navigation, and Timing Advisory Board, April, https://www.gps.gov/multimedia/presentations/1997-2004/2004-04-ieee/5-ImprovingTheGPSL1Signal.pdf.

- L2 is centered at 1227.6 MHz and carries civilian (L2 C) and military (L2 P(Y), 20.46 MHz bandwidth, and L2 M, 30.69 MHz bandwidth) signals.
- L5 is centered at 1176.45 MHz and carries a civilian signal (L5 and Q5, 20.46 MHz bandwidth).

A new civilian signal on L1 (termed L1 C) is currently operating on five satellites. It occupies 30.69 MHz, with a 4 MHz wide main lobe and significant side lobes out to 14 MHz.

For typical users, a minimum of four GPS satellites are needed to obtain a navigation fix; precision and accuracy can be improved if the user enjoys reception from more than four satellites. The navigation fix is based on measured pseudo-ranges, which are determined by subtracting the transmission time from the reception time and multiplying this difference by the speed of light in a vacuum. Pseudo-range is a combination of true range, transmitter and receiver clock offset effects, effects from signal slowing in the Earth's ionosphere and neutral atmosphere, and multipath-induced spreading of the signal arrival time. A navigation fix first corrects for each satellite clock offset by using SV transmitted clock calibration data and for ionosphere and neutral atmosphere delay by using models. The receiver clock offset is estimated along with the navigation coordinates, which is why four or more satellites are needed.

The actual received signal power depends on many conditions at the observing site, including the elevation angle of the satellite above the horizon, obstructions, and multi-path interference.

1.2.2 Military GPS Evolution and Policy

On May 23, 1983, when the Navstar GPS satellite production contract was signed, the payload provided wideband encrypted signals on L1 (1575.46 MHz) and L2 (1227.6 MHz) and the unencrypted coarse/acquisition (C/A) signal on L1. Early military GPS receivers were not capable of acquiring the encrypted wideband signals directly given the limits of receiver electronics in those days. Those receivers first acquired GPS through the narrowband, unencrypted L1 C/A signal, which could be easily jammed or spoofed.

The Soviet Union shot down Korean Airlines (KAL) Flight 007 on September 1, 1983. On September 16, 1983, President Ronald Reagan's Deputy Press Secretary Larry Speakes announced, "The President has determined that the United States is prepared to make available to civilian aircraft the facilities of its Global Positioning System when it becomes operational in 1988."

Ultimately, the L1 C/A and P(Y) and L2 P(Y) became the Precise Positioning Service (PPS). The unencrypted L1 C/A was designated the Standard Positioning Service (SPS) to fulfill President Reagan's promise to civil aviation and for civil and commercial use in general.

As technology advanced, the GPS Joint Program Office began development of a Selective Availability Anti-Spoofing Module (SAASM) in the mid-1990s, which enabled military GPS receivers to acquire L1 P(Y) or L2 P(Y) directly and securely.

Chairman of the Joint Chiefs of Staff Instruction (CJCSI) 6140-01A, "Navstar Global Positioning System (GPS) Selective Availability Anti-Spoofing Module (SAASM) Requirements," dated March 31, 2004, mandated GPS PPS SAASM-based military receivers for all new procurements after October 1, 2006.

In response to concerns about potential electronic warfare threats, Vice President Al Gore announced plans to modernize military GPS at a March 29, 1996, congressional hearing. The U.S. Department of Defense (DoD) started development of a new military satellite signal, M-Code, along with upgrades to the ground control system to manage the new military signal, and military GPS user equipment with upgraded cybersecurity and antijam capabilities in the summer of 1997.

The first GPS satellite to transmit M-Code was launched on September 26, 2005, and it is still operational. According to the Air Force Fiscal Year (FY) 2023 Budget Submission, Initial Operational Capability (IOC) of Operational Control System (OCX) Blocks

1 and 2, which allow the GPS ground control system to fully operate M-Code, is now planned for the first quarter of FY 2024 and M-Code Military GPS User Equipment (MGUE) Increment 1 Lead Platform Integration and Test is scheduled for completion in the second quarter of FY 2025 with Enterprise M-Code Positioning, Navigation, and Timing (PNT) IOC declaration planned shortly after.

With respect to DoD use of non-U.S. military PNT, the relevant DoD instruction states:

> a. Reliance on civil, commercial, or foreign sources as the primary means of obtaining PNT information for combat, combat support, or combat service support operations is not authorized without a waiver in accordance with CJCS Instruction (CJCSI) 6130.01G. These systems may be utilized as complementary sources, subject to successful NAVWAR compliance determination.
>
> b. Use of civil, commercial, or foreign sources to obtain PNT information for non-combat operations is authorized, subject to successful NAVWAR compliance determination.[13]

1.2.3 GPS Satellites and Transmitted Signal Structure

GPS signals are broadcast in three frequency bands. The L1 signal is transmitted in the band centered on 1575.42 MHz, as depicted in Figure 1-4. The L2 band is centered on 1227.6 MHz. Last, the L5 band is centered on 1176.45. Because Ligado's downlink is between 1526 and 1536 MHz, its impact is primarily relevant to the L1 band.

The actual L1 signal for Block II satellites[14] is

$$s(t) = A_c C(t) D(t) \sin(\omega_c t + \theta_1) + A_p P_Y(t) D(t) \cos(\omega_c t + \theta_2) + A_m M(t) \cos(\omega_c t + \theta_2),$$

where $C(t)$ is the coarse acquisition code; $D(t)$ is the navigation signal consisting of clock corrections, ephemeris, and almanac, amounting to about 37,500 bits of data broadcast once every 12.5 minutes, with the most important data constituting less than 1,500 bits and broadcast once every 30 seconds; $P_Y(t)$ is the military precision ranging code (an encrypted version of the known $P(t)$); and $M(t)$ is a newer encrypted military code.

The C/A signal (the first term in $s(t)$ above) is composed of a digital navigation data stream ($D(t)$ above) broadcast at 50 bits per second multiplied by a "chipping" code

---

[13] U.S. Department of Defense, 2020, "DoD Instruction 4650.08: Positioning, Navigation, and Timing (PNT) and Navigation Warfare (NAVWAR)," Change 1, December 30, Washington, DC, https://www.esd.whs.mil/Portals/54/Documents/DD/issuances/dodi/465008p.pdf?ver=M9B6zSt5uWSeDoPwocp_RQ%3D%3D.

[14] Block III satellites also broadcast the L1C signal.

($C(t)$) with a frequency of 1023 million chips per second). The navigation data stream and chipping code are synchronous. The binary symbols in the time series $C(t)D(t)$ are modulated into the 1575.42 MHz L1 carrier using Binary Phase Shift Keying (BPSK). The chipping code provides noise immunity so that the navigation code can be reliably received and demodulated at GPS receivers; in addition, the chipping code is SVN dependent, so individual SV signal streams can be distinguished. Most importantly, the chipping code provides the absolute ranging capability that enables the determination of pseudo-range, which is the fundamental measurement that underpins standard GPS radionavigation.

The legacy L1 C/A navigation message consists of 37,500 message bits and is transmitted at 50 bits per second. This message is composed of 25 pages of 30 seconds duration, and thus takes 12.5 minutes to transmit. Every page is further subdivided into five 6-second sub-frames; every sub-frame consists of ten 30-bit words.

The content of the navigation message is as follows. Each sub-frame begins with the telemetry word (TLM), for synchronization. It is followed by the transference word (HOW), which provides time information that allows the receiver to acquire the week-long P(Y)-code segment. Sub-frame 1 contains the satellite clock information and information about satellite health. Sub-frames 2 and 3 contain satellite ephemeris. Sub-frame 4 provides ionospheric model parameters, Universal Coordinated Time (UTC) information, and part of the almanac. It also indicates whether anti-spoofing, A/S, is activated. Sub-frame 5 contains the constellation status and data from the almanac, allowing a receiver to more rapidly acquire the satellites. Twenty-five frames are required to complete the almanac.

Sub-frames 1, 2, and 3 are transmitted as part of every frame, and the content of sub-frames 4 and 5 is common across all satellites. Thus, almanac data for all satellites in orbit can be obtained from a single tracked satellite.

The legacy L1 C/A Navigation Message—time parameters and clock corrections, ephemeris parameters, almanac, as noted above, along with Service Parameters with ionospheric model parameters and satellite health information—allows the position of any satellite in the constellation to be calculated. The ephemeris and clocks parameters are normally updated every 2 hours, and the almanac is updated at least every 6 days.

When a receiver successfully demodulates the ephemeris and clock correction parts of the navigation message from four separate GPS satellites, it can use the message contents to calculate the position of the receiver (i.e., latitude, longitude, and height above the Earth's sea-level ellipsoid). This calculation generally provides positioning accuracy of no worse than 5 m, but accuracy can be further improved by data from the Wide Area Augmentation System (WAAS). High-precision receivers use carrier phase observations and better models to achieve centimeter accuracy and are discussed further in Section 1.2.5. Averaging over time can often improve accuracy to the order of millimeters for high-precision receivers.

The L1 C/A chipping code is a 1023-bit pseudo-random subset of a Gold code (which can be calculated using two linear feedback shift registers) particular to an SVN. The chips can be denoted as $c_i = \pm 1$, $i = 1, 2, ..., 1023$, and a navigation message bit as $b = \pm 1$. After chipping, b is represented by 1023 bits of the form $m_i = bc_i$, where m_i is BPSK modulated onto the 1575.42 L1 carrier, as indicated in Figure 1-5. A receiver can replicate the chipping code given the SVN and can align its replica chip streams by cross-correlating with the received values m_i. The receiver may not know the visible SVNs a priori and so it guesses the SVN and selects the candidate SVNs with the best signal autocorrelation among the trial SVNs. This is part of the process of code acquisition. Chipping, or direct sequence spread spectrum (DSSS), and the associated de-spreading computations in a receiver via code wipe-off greatly improves the effective signal-to-noise ratio (SNR) of a GPS signal, providing robust noise immunity.

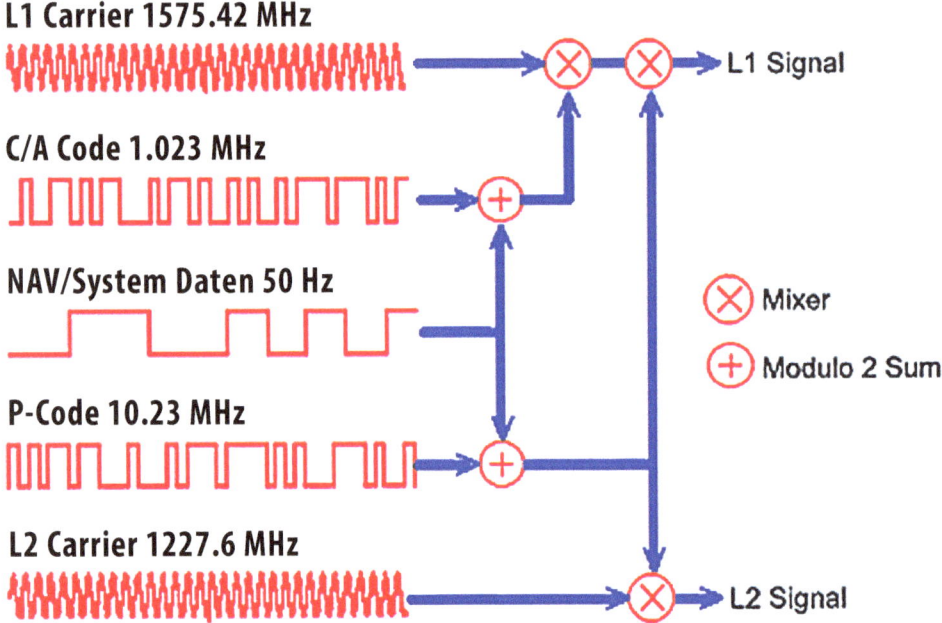

FIGURE 1-5 Modulated GPS signals for the original system design.
SOURCE: P.H. Dana, 1994, "The GPS Overview," The Geographer's Craft Project, University of Colorado Boulder, https://foote.geography.uconn.edu/gcraft/notes/gps/gps_f.html.

Robust noise immunity is critical for GPS performance because the observed power levels at the surface of the Earth are on the order of −160 dBW.[15] Given a device antenna gain of 4 dBi, the received power, from a satellite near zenith, is about −155 dBW, while the noise power over the 2.046 MHz signal bandwidth is about −140 dBW. Of course, signal power degrades as elevation decreases, but there are usually four or more satellites visible at an elevation of 30 degrees or higher. More detailed signal models that incorporate the effect of satellite elevation and ionospheric interference can be found in Misra and Enge.[16] A key observation is that the thermal noise power at a receiver on Earth is higher than signal power.

The legacy L1 C/A signal pseudo-random noise (PRN) code, modulation technique, data rates, data format, and information content were developed in the early 1970s based on the receiver and information system technology of that era as well as projections of how GPS military receivers might operate given those limitations. Twenty-five years later, in the mid-1990s, when DoD initiated the military GPS modernization program, to include M-Code, the GPS Joint Program Office (JPO) realized that receiver technology and connectivity had progressed to the point that a static 37,500-bit message, transmitted at 50 bits per second, and taking 12.5 minutes to receive, was inefficient, was militarily ineffective, and made military receivers vulnerable to exploitation because L1 C/A is unencrypted. M-Code adopted a signal structure that employed modernized modulation techniques, data rates, data format, information content, and flexible messages that were 1,000 times shorter and could be changed by the GPS ground control system so that receivers could receive the information needed much faster. The first M-Code Interface Control Document describing these features was formally published on August 24, 2001. In parallel with its work on M-Code, the GPS JPO proposed using new, more robust PRN codes and different data rates as well as shorter, flexible data messages on L2 similar to M-Code, naming the new signal L2C. On August 17, 2001, the U.S. Department of Transportation (DOT) agreed to this approach and requested the GPS JPO begin implementation of L2C. Shorter, flexible data messages and more robust PRN codes were previously implemented on L5 as well as L1C. Unlike L1 C/A, L2C, L1C, and L5 also have data-less pilot channels that increase performance and interference protection. Nevertheless, a great majority of civilian receivers, including all certified aviation receivers,[17] still rely on the L1 C/A code and its 50 Hz navigation data

---

[15] See Table 3-Va in Interface Control Working Group, 2022, "NAVSTAR GPS Space Segment/Navigation User Interfaces," IS-GPS-200 Revision M, April 13, approved May 21, El Segundo, CA: SAIC, https://www.gps.gov/technical/icwg/IS-GPS-200M.pdf.

[16] P. Misra and P. Enge, 2011, *Global Positioning System: Signals, Measurements, and Performance*, Lincoln, MA: Ganga-Jamuna Press.

[17] RTCA, Inc., 2006, "Minimum Operational Performance Standards for Global Positioning System/Wide Area Augmentation System Airborne Equipment," RTCA-DO-229-Revision D, Washington, DC, December 13.

stream and cannot operate without this signal and its navigation data. Complete definitions of the GPS downlink Navigation Signals can be found in the applicable GPS interface specification documents.[18]

Neither L1C, L2C, L5, nor M-Code have been declared operational by the U.S. Space Force. Twenty-four operational satellites are typically required for that to occur. As of April 12, 2022, there are 4 GPS satellites broadcasting L1C, 23 broadcasting L2C, 16 broadcasting L5, and 23 broadcasting M-Code. In its FY 2023 budget request to Congress, U.S. Space Force shows M-Code, L2C, and L5 IOC as the first quarter of FY 2024.

1.2.4 GPS Receivers

The design of a notional GPS receiver is depicted in Figure 1-6. The radio frequency (RF) or analog stage of the GPS receiver consists of the blocks from the antenna through the analog-to-digital (A/D) converter, and the digital stage begins after the A/D. In practice, receivers can be much more complicated, with multiple intermediate frequency (IF) stages. In addition, some receivers perform additional filtering after the signal is digitized.

GPS receivers are specialized. There are general location/navigation receivers (GLNs) such as those in a portable GPS unit, cellular receivers (CELs) such those in a cell phone, and timing (TIM) and high-precision (HP) receivers that use carrier phase measurements to improve accuracy. HP receivers come in several varieties. Those operating

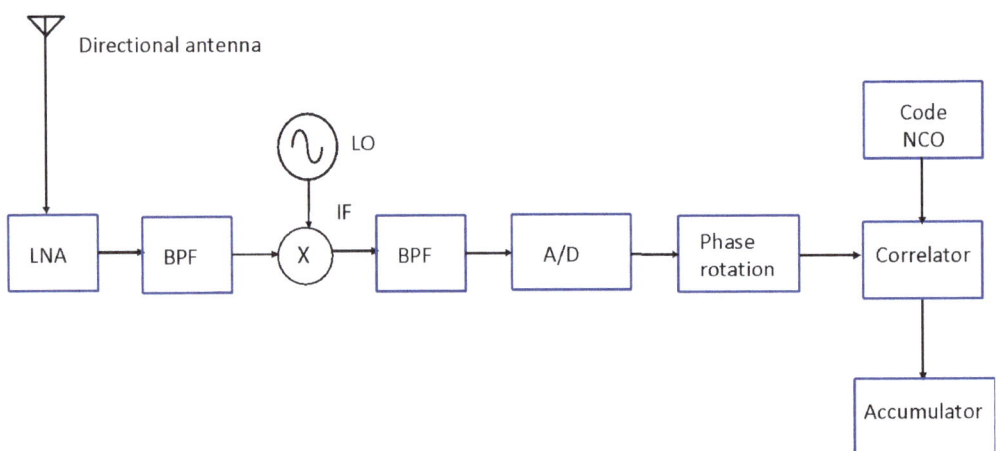

FIGURE 1-6 Notional Global Positioning System (GPS) receiver.
NOTE: A/D, analog-to-digital converter; BPF, band-pass filter; IF, intermediate frequency; LNA, low noise amplifier; LO, local oscillator; NCO, numerically controlled oscillator; the circle with an "x" is a mixer.

---

[18] L1 P(Y), L2 P(Y), L1 C/A and L2C in IS-GPS-200; L1 and L2 M-Code in ICD-GPS-700; L5 in IS-GPS-705; and L1C in IS-GPS-800.

in real time use software that can produce the needed high precision in real time. To achieve very high accuracy, the same equipment can be used by scientific users that might opt to analyze these data at longer cadences (e.g., hourly or once per day).

The filters have the important job of selecting, to the extent possible, received power only in the band of interest (1560–1591 MHz for L1), so as to receive only the signal that is desired to be processed. For example, for L1 C/A processing, a receiver might employ a narrow filter with a bandpass of about 2 MHz; by contrast, M-code processing might employ a filter with a 20–30 MHz bandpass. A perfect filter would eliminate all power outside the band of interest or "co-channel." Filters are not perfect; some power in nearby frequency bands makes its way to the final processed digitized signal, although in attenuated form. In principle, with good enough filters, signals away from the co-channel in frequency do not interfere with receiver performance. A well-designed receiver substantially eliminates power from bands outside the co-channel, presenting a "clean signal" for later signal processing.

Another possible issue caused by out-of-band power is desensitization of the initial LNA owing to nonlinear compression if it receives too much input power. The risk for this is elevated in a multiband receiver where a single antenna and LNA have been designed to receive, say, L1, L2, L5, Galileo, and GLONASS signals. If the Ligado signal were to send the LNA of such a receiver into compression, then the remedy would be to add a notch filter between the antenna and the LNA in order to attenuate the Ligado band. This filter would reduce the sensitivity and SNR of the affected receiver, but it could preserve the receiver's ability to operate in the presence of Ligado signals.

As will be discussed further below, under the FCC order, co-channel interference from Ligado mobile terminals is limited to −105 dBW/MHz throughout the co-channel or band of interest. Ligado interference in adjacent channel is limited to −85 dBW/MHz near the co-channel. With appropriate filtering, the adjacent power, *at the license specified levels, in the vast majority of existing GPS receivers does not have a material adverse effect on receiver circuits processing GPS signals*; in addition, practical receiver performance characteristics of the sort recommended in later sections of this report can ensure that this is the case.[19]

The increase in effective noise density is often calculated as $(I + N_0)/N_0$, where N_0 is the ambient noise power and I is the interference power in a frequency range experienced at the receiver from an unwanted signal. A 1 dB C/N_0 signal degradation

---

[19] The FCC does not generally specify expected receiver characteristics calculated to eliminate the possibility of adjacent (or distant) band interference levels. However, certified receivers, such as those used in aviation, do have such associated requirements. Past FCC studies have recommended that the FCC publish such receiver performance characteristics, as the committee does later in this report.

corresponds to $(I + N_0)/N_0$ of 1.259, meaning that the unwanted signal contributes an additional 26 percent over the noise floor to the effective noise power. A 3 dB C/N_0 signal degradation corresponds to $(I + N_0)/N_0$ of about 2, meaning the unwanted signal contributes an additional 100 percent. A 5 dB C/N_0 signal degradation corresponds to $(I + N_0)/N_0$ of about 3.2, meaning the unwanted signal contributes an additional 220 percent.

Under test conditions, many GPS CEL receivers tolerate[20] noise equivalent to an 8 dB C/N_0 signal degradation. Almost all GPS receivers operating in open-sky conditions, including most HP receivers, tolerate noise equivalent to a 4 dB C/N_0 signal degradation, or six times the equivalent of a 1 dB C/N_0 signal degradation before the onset of harmful interference.

The additional interference power in a frequency band (*I* above) at a receiver is dependent not only on the power of an emitter, the directionality of the transmitting and receiving antennas, and the "sharpness" of an emitter's transmission filter, but also on the distance between the emitter and the transmitter. Because received power declines as the square of the distance between the emitter and the receiver for a free-space model, and by a power as high as 3.5 for other models, this has a more profound effect on interference than transmission power. For example, using the free space loss model, if a receiver experiences a 100 percent increase in interference power (over the noise floor) at 3 m from an emitter, it will experience only a 25 percent increase at 6 m of separation and a 1 percent increase in interference power over the noise floor at 30 m of separation.

C/N_0, as presented at an antenna, is difficult to measure directly. Rather, the laboratory experiments conducted to evaluate interference available to the committee employ *receiver reported* C/N_0 to estimate the effect of Ligado emissions. Receiver reported values are subject to the device's opaque, proprietary algorithm and could respond non-linearly, logarithmically, or otherwise. This is often referred to as "effective C/N_0." Effective C/N_0 is an explicit measure of the result of interference on GPS receiver functions. It is for this reason that effective C/N_0 is the standard figure of merit used by experts in the field to measure receiver health. Receiver reported C/N_0 values are, of course, receiver dependent. These figures include receiver noise and processing noise as well as adjacent power caused by RF front-end limitations. Using receiver reported C/N_0 greatly complicates the analysis of noise from Ligado emitters, because it requires careful testing of each unit. Nevertheless, almost all of the available studies use receiver reported C/N_0 degradation as the figure of merit for the onset of interference.

---

[20] "Tolerate" here means it suffers no significant degradation in observable performance like position accuracy, acquisition time, or loss of tracking.

GPS acquisition refers to the receiver's process of aligning its replica chipping code to the SV transmission, determining the Doppler shift of the received carrier signal, or "locking" onto the signaled bit stream. To decode the navigation signal, the receiver must maintain alignment with the C/A code (or the P(Y) code in a military receiver). Acquisition is affected first and foremost by the effective SNR of the transmitted signal at the receiver, which, in turn, affects the quality of the signal presented to the A/D by the RF front end of a GPS receiver. This is a rather receiver-dependent process. If, for example, a receiver admits too much power from adjacent channels, the signals will be much noisier and harder to demodulate and to track.

After digitization, additional signal processing is performed. For example, Doppler frequency shifts caused by the relative motion between the SV and the receiver are corrected. The signal is demodulated, and it is here that the C/A code tracking and carrier phase tracking is performed. This signal processing for the navigation/timing solutions is described briefly below. How interference can adversely affect this processing is discussed later in this report.

Most GPS users rely on pseudo-range observables and the broadcast orbits and clocks corrections. A simplified observable equation for a pseudo-range signal received on the L1 frequency $p_1(t_r)$ can be defined as:

$$p_1(t_r) = \| \vec{X}^s(t^s) - \vec{X}_r(t_r) \| + c\delta_r - c\delta^s + \rho_t + \rho_{i1} + \rho_{m1} + \varepsilon_{p1}.$$

The first (and largest) term here is true range derived from the three-dimensional coordinates of the satellite, $\vec{X}^s$, when the signal was sent, t^s, and the receiver position when it was received, $\vec{X}_r(t_r)$. Navigation users rely on the GPS ephemeris provided on the signal itself to model $\vec{X}^s(t^s)$, which is currently approximately 1 m in precision in a radial sense. The satellite clock offset (δ^s) is also provided in the broadcast ephemeris and is of similar precision. Fairly crude models can be used to remove most of the tropospheric ρ_t and ionospheric ρ_{i1} effects. For simplicity, the multi-path effect, ρ_{m1}, is not discussed here. The measurement noise term is ε_{p1} (1–2 m). This leaves a navigation user with four unknowns: the receiver's three-dimensional position $\vec{X}_r$ and the receiver clock offset (δ_r). If pseudo-range observations from four or more satellites are available, these four terms can be estimated.

From one SV, a GPS receiver A/D samples the signal

$$r(t) = \sqrt{S}D(t - \tau)C(t - \tau) \cos(2\pi(f_{IF} + f_D)t + \delta\theta) + n(t),$$

where S is the mixed signal power, f_{IF} is the IF frequency, f_D is the Doppler frequency shift, $\delta\theta$ is the carrier phase offset, and $n(t)$ is the noise. τ is the signal delay caused by the speed of light signal transport. The received signal is mixed with the receiver's local oscillator signal $\sqrt{2}\cos(2\pi(f_{L1} - f_{IF})t + \theta_{IF})$. θ_{IF} is randomly distributed with respect to the incoming signal phase. The GPS receiver's job is to get an accurate estimate of $(\tau, f_D, \delta\theta)$.

A GPS receiver generates two reference signals:

$$I(t) = 2\cos(2\pi(f_{IF} + \hat{f}_D)t + \delta\hat{\theta})C(t - \hat{\tau}),$$
$$I(t) = -2\sin(2\pi(f_{IF} + \hat{f}_D)t + \delta\hat{\theta})C(t - \hat{\tau}),$$

where $\hat{\tau}$ is the receiver estimated signal delay, $\hat{f}_D$ is the receiver estimated Doppler shift, and $\delta\hat{\theta}$ is the receiver estimated phase. These signals are, respectively, the in-phase and quadrature reference signals and are often represented by the analytic signal $I(t) + jQ(t)$. These two signals are multiplied by the received signal $r(t)$, and the resulting samples are accumulated into in-phase and quadrature sums. This feedback of such sums allows the receiver to adjust its estimates of $\hat{\tau}$, $\hat{f}_D$, and $\delta\hat{\theta}$ so that they closely match the true values of, respectively, τ, f_D, and $\delta\theta$. A good GPS receiver accurately tracks both C/A code alignment and carrier phase alignment. These are the two "tracking loops" in a GPS receiver. Both loops employ numerically controlled oscillators (NCOs).

Accuracy can be improved by way of carrier-phase enhancement. This technique uses the L1 carrier wave, which has a period 1/1540th of the duration of the L1 C/A chipping code rate, providing an additional clock. The code phase error in the normal GPS causes 2–3 m error in position accuracy. Carrier phase enhancement reduces this to 1–2 cm (0.4–0.8 in). Carrier phase enhancement is complicated by the ambiguity caused by full cycle (2π) shifts of a signal. Carrier phase enhancement is required in high-precision receivers.

1.2.5 High-Precision GPS Users and Considerations

The previous section introduced the pseudo-range observable equation on the L1 frequency and outlined a simple description of how the receiver uses these observations to estimate position and timing. High-precision GPS users rely on the same GPS signals, but use carrier phase observations (ϕ_1) instead of pseudo-range:

$$\lambda_1 \phi_1(t_r) = \| \vec{X}^s(t^s) - \vec{X}_r(t_r) \| + c\delta_r - c\delta^s + \rho_t - \rho_{i1} + \rho_{m\phi1} + N\lambda_1 + \varepsilon_{\phi1}.$$

There are three main differences between the pseudo-range and carrier phase equations:

1. There is an extra term (*N*), the carrier phase bias, which is scaled by the signal wavelength λ_1. It remains the same for a given satellite pass as long as the receiver maintains lock on the signal. It must be estimated. High-precision GPS users are concerned about loss of lock because each loss of lock requires that this term be re-estimated. Needing to do so severely degrades their position solutions.
2. The ionospheric delay ρ_{i1} has the opposite sign as the pseudo-range ionosphere term.
3. The carrier phase measurement uncertainty ($\varepsilon_{\phi 1}$) is nearly two orders of magnitude smaller (1–2 cm) than the pseudo-range measurement uncertainty.

Either pseudo-range or carrier phase observations can be used to estimate the receiver position ($\vec{X}_r$). However, one cannot directly achieve cm-level positioning using carrier phase data with the broadcast orbits and clocks. Early on, GPS users realized that this problem could be mitigated by using two GPS receivers and estimating "differential" baselines rather than "absolute positions." This practice is called differential positioning. It is the underpinning of today's real-time kinematic (RTK) applications. The location of a receiver can be found by combining its data in software with data from the second "base" station. The difference of the two stations' carrier phase observations is relatively insensitive to satellite orbit errors and atmospheric delay errors, meaning even the crude broadcast orbits can be used. The clock effects are completely removed in this way. Initially users of this technique deployed their own base stations, but today networks of base stations are operated both by the government and commercial entities.

The users of high-precision GPS can be loosely grouped in three categories:

- *Surveyors* make critical contributions to the national infrastructure and define legal land boundaries. They were early adopters of GPS and are also significant users of Galileo, BeiDou, and GLONASS signals. They primarily use the differential GPS technique, and the GPS receiver is assumed to be stationary. These users require *centimeter-level accuracy*. This community has especially benefited from software that allows them to quickly (several minutes) estimate their three-dimensional coordinates. The ability to measure land boundaries with this precision and at this speed has obvious commercial implications. Although modern surveyors would not be worried about downloading orbits on the GPS signals (e.g., cold starts), they cannot easily tolerate interference leading to loss of lock.

- *RTK users* employ similar principles as surveyors, but as the acronym suggests, they measure differential baseline parameters in real time. As with the other user groups, RTK users cannot tolerate interference so severe that it leads to simultaneous loss of lock across many channels. RTK is broadly used in the United States, for example in agribusiness, which uses it to locate farming equipment, and the construction industry, which uses it at building sites. They have similar accuracy requirements as surveyors, *centimeter level*, with the proviso that there are some applications with extremely short baselines (e.g., two antennas on a single vehicle) that could have more stringent requirements.

- *Scientific users of high-precision GPS* are mostly geodesists and geophysicists. These users study scientific issues (measuring plate tectonics, determining group deformation caused by earthquakes or volcanic inflation/deflation) that require three-dimensional positions (daily average) with *a horizontal precision of around several millimeters*. The atmospheric delays on GPS signals (and on those of any other Global Navigation Satellite System [GNSS], such as Galileo, GLONASS, and BeiDou) are used for both climate and forecasting studies by meteorologists and space weather scientists. GPS data are also now used to provide early warnings of such hazards as tsunamis and earthquakes.

    Scientific users of GPS signals ignore all models transmitted by DoD (orbits, clocks, ionospheric models) or by receiver manufacturers that sell similar products to their users. There are long-lived geodetic communities (since 1994) that operate GPS (and GNSS) instruments around the world and provide precise orbits and clocks.[21] They require the most accurate models for Earth's orientation in space, the GPS spacecraft, and ground instrumentation (antenna phase centers and gain patterns primarily). They have demanded that receiver manufacturers provide multi-path suppressing antennas with (azimuthally) homogeneous gain patterns. Data used in this community were initially downloaded from Internet sites, but increasingly these data are streamed. Because these users run their sites continuously, cold starts are for the most part irrelevant, but they cannot tolerate interference leading to loss of lock.

---

[21] See International GNSS Service, 2020, "Homepage," https://igs.org.

1.3 FCC ORDER 20-48

On April 19, 2020, the FCC adopted FCC Order 20-48,[22] authorizing Ligado Networks, LLC, to move forward with deployment and operation of a low-power terrestrial nationwide network in the 1526–1536 MHz, 1627.5–1637.5 MHz, and 1646.5–1656.5 MHz bands.

1.3.1 The FCC's Analysis in the Order

The FCC order concludes that there are "significant public interest benefits associated with Ligado's proposed ATC network deployment" as detailed in Ligado's amended license modification. After considering "the concerns raised regarding potential harmful interference to adjacent band operations," including general location and navigation devices, certified aviation GPS devices, non-certified GPS receivers, other U.S. government devices, as well as MSS operations, the FCC order concludes that, subject to the various conditions it imposes on Ligado's ATC authority, granting the license "will promote the efficient and effective use of our nation's spectrum resources." The FCC thus granted the license subject to certain conditions, detailed in Section 1.3.2 below.

Ligado entered into a number of so-called co-existence agreements with several equipment manufacturers (Garmin, Deere, Trimble, NovAtel, Topcon, and Septentrio), representing a major share of the market for general location and navigation, high-precision location and navigation, and non-certified aviation location devices. At various times throughout the long process that has led to this approval, several of these companies provided input to the FCC indicating the potential for harmful interference and objecting to the granting of the then-current Ligado application. Bilateral discussion with Ligado led to a series of commitments by Ligado to each company based on specific emission-level requirements and led to each of the co-existence agreements. The analysis presented in the FCC's final order relies on these agreements to indicate support for approval of the license:

> We find that the co-existence agreements that Ligado has reached with each of the manufacturers discussed above are significant and support the Commission's finding that technical and operational solutions to address concerns about harmful interference to GPS receivers have been developed and that, with appropriate notifications and other conditions (discussed below), these solutions can address other potential

---

[22] In the Matter of LightSquared Technical Working Group Report (IB Docket No. 11-109), LightSquared License Modification Application (IB Docket No. 12-340), New LightSquared License Modification Applications (IB Docket No. 11-109; IB Docket No. 12-340), Ligado Amendment to License Modification Applications (IB Docket No. 11-109).

Introduction and Background

harmful interference concerns as Ligado's terrestrial network operations are rolled out over time.[23,24]

In reaching its conclusion to grant the license, the order relies on its determination of whether there will be interference and whether such interference meets the definition of "Harmful Interference." These definitions are crucial to the FCC's conclusions regarding authorizing the Ligado system.

The analysis in the order presents the two different measures of interference that are at the heart of the first task that this committee was asked to address. The first of these is an operational definition based on the measured performance of a device under test (DUT), that is, at what emission levels is a device under test no longer able to provide the measurements for which it is intended? Those measurements include position and/or timing accuracy, but also may include other key performance indicators (KPIs), such as the capability and timing to (re)acquire, lock, and maintain signal integrity. As an operational measurement of interference, this method can be used only by actual testing of receivers under consideration. The second method to determine interference is to use an "interference tolerance mask." Typically, this means answering the question of whether a receiver sees a change in the carrier-to-noise density ratio (C/N_0) in the presence of the emitted power from the source under question. It is often the case that a -1 dB $\Delta C/N_0$ is used as the threshold for interference. As noted previously, these are both measurements of interference and must be interpreted based on the FCC's definition of Harmful Interference:

> Interference which endangers the functioning of a radionavigation service or of other safety services or seriously degrades, obstructs, or repeatedly interrupts a radiocommunication service operating in accordance with [the ITU] Radio Regulations.

Therefore, Section 1.4 of this report provides background and a detailed analysis on the general subject as well as the difference between the technical definition of interference and the operational definition of the harmful interference. This grounding is important as a precursor to the committee's responses to the questions one and two of the study charter.

---

[23] FCC Order 20-48, paragraph 34, p. 19.

[24] It should be noted that in their presentations to the committee, Garmin and Trimble commented that these legal agreements should not be construed as inferring that all technical issues about potential interference had been resolved to the satisfaction of Garmin and Trimble. Most, if not all, of the equipment manufacturers listed here have filed statements with the FCC to the effect that none of the agreements they signed were on "coexistence" but rather lawsuit settlement agreements. See, for example, the section "The Commission Overstates and Mischaracterizes the Relevance of Ligado's Agreements with GPS Manufacturers" of the Trimble Petition for Reconsideration filed with the FCC on May 22, 2020.

The FCC order reviews the various test reports performed over the course of the proceedings, including a series of reports from Roberson and Associates (*Results of GPS and Adjacent Band Co-Existence Study*, May 9, 2016; *Final Report: GPS and Adjacent Band Co-Existence Study*, June 10, 2016; collectively the RAA Reports); a GPS device testing report commissioned by Ligado and performed by the National Advanced Spectrum and Communications Test Network (NASCTN), titled *LTE Impacts on GPS Final Report* (the NASCTN Report); and an April 2018 report from the DOT, titled *Global Positioning System (GPS) Adjacent Band Compatibility Assessment Final Report* (the DOT ABC Report).

The order reviews the details of these reports and presents the findings of those who advocated for or against approval of the license. In particular, the order notes that at a late stage in the Commission's proceedings it received a letter[25] from the NTIA, including an Air Force memorandum to the NTIA's Interdepartmental Radio Advisory Committee (IRAC). This letter cites a March 2018 report[26] from the National Space-Based Positioning, Navigation, and Timing Systems Engineering Forum (NPEF), which provided an assessment of testing methodologies that had been used in these three reports cited above. The NTIA's letter stated its (the NTIA's) belief that the Commission "cannot reasonably reach" a conclusion that harmful interference concerns have been resolved.

Section 2.2 of this report, in providing the committee's response to Task 2 of its charter, discusses the results of the various studies noted above.

In basing its decision to authorize the Ligado license, the FCC stated that consistent with past practices, the technical measurement of a change in carrier-to-noise density does not provide an appropriate connection to the actual impact of interference in performance (and ultimately to harmful interference). Its conclusion rests primarily on the performance-based assessments as detailed in the reports referenced above, on the lack of data-driven objections from potential affected parties, the various co-existence agreements, and Ligado's commitments to remediation.

1.3.2 Requirements and Conditions on the Authorization

FCC Order 20-48 approved the Ligado amended application to operate subject to numerous requirements and conditions, which are summarized in Box 1-1.

---

[25] Letter from Douglas W. Kinkoph, NTIA to Ajit Pai, Chairman, FCC, IB Docket No. 11-109, p. 2 (filed April 10, 2020).

[26] NPEF, 2018, "Assessment to Identify Gaps in Testing of Adjacent Band Interference to the Global Positioning System (GPS) L1 Band," https://www.gps.gov/spectrum/ABC/2018-03-NPEF-gap-analysis.pdf.

> **BOX 1-1 Requirement and Conditions in FCC Order 20-48**
>
> 1. In-Band Emissions
>   a. Downlink power emitted from Ligado Ancillary Terrestrial Component (ATC) base stations in the 1526–1536 MHz band must not exceed an effective isotropic radiated power (EIRP) of 9.8 dBW measured with a +/−45-degree cross-polarized antenna. This places a minimum station to station separation distance of 433 m.
>   b. Uplink power [from a user to a fixed tower] in the 1627.5–1632.5 MHz band must initially not exceed an EIRP level defined as a linear ramp starting at −31 dBW at 1627.5 MHz and ending at −7 dBW at 1632.5 MHz. After a period of 5 years, the uplink power in this band will be subject to the −7 dBW across the entire band.
>   c. Uplink power in the 1646.5–1656.5 MHz band must not exceed an EIRP level of −7 dBW.
>   d. Ligado is not permitted to operate using its ATC authority in its mobile satellite services (MSS) downlink spectrum from 1545–1555 MHz.
>   e. Ligado is prohibited from operating any ATC base station antenna in its lower downlink band, 1526–1536 MHz, at a location less than 250 feet laterally or less than 30 feet below an obstacle clearance surface established by the Federal Aviation Administration (FAA) under 14 CFR Part 77 (and its implementing orders and decisions).
> 2. Out-of-Band Emissions
>   a. The EIRP from a Ligado ATC mobile terminal must not exceed a ceiling defined by the following:
>   i. −67 dBW/4kHz at 1627.5 MHz,
>   ii. Linear interpolation from −67 dBW/4kHz at 1627.5 MHz to −100 dBW/MHz at 1610 MHz in the 1627.5–1610 MHz frequency range,
>   iii. Linear interpolation from −100 dBW/MHz at 1610 MHz to −105 dBW/MHz at 1608 MHz in the 1610–1608 MHz frequency range; −105 dBW/MHz in the 1541–1608 MHz frequency range, and
>   iv. −58 dBW/4 kHz at a 1 MHz offset beyond the edges of the MSS frequency assignment at 1646.5–1656.5 MHz.
>   b. The EIRP from a Ligado ATC mobile terminal's discrete emissions must not exceed a ceiling defined by the following:
>   i. −44 dBW/700 Hz at 1625 MHz,
>   ii. Linear interpolation from −44 dBW/700 Hz at 1625 MHz to −110 dBW/700 Hz at 1610 MHz,
>   iii. Linear interpolation from −110 dBW/700 Hz at 1610 MHz to −115 dBW/700 Hz at 1608 MHz,
>   iv. −115 dBW/700 Hz in the 1608–1559 MHz frequency range,
>   v. −132 dBW/2 kHz in the 1559–1541 MHz frequency range.
>   c. The EIRP from a Ligado ATC base station's out-of-band emissions must not exceed a ceiling defined by the following:
>   i. −85 dBW/MHz in the 1541–1559 MHz and 1610–1650 MHz frequency ranges,
>   ii. −100 dBW/MHz in the 1559–1610 MHz frequency range.
>   d. The EIRP from a Ligado ATC base station's discrete emissions must not exceed a ceiling defined by the following:
>   i. −112 dBW/2 kHz in the 1541–1559 MHz frequency range,
>   ii. −110 dBW/700 Hz in the 1559–1610 MHz frequency range,
>   iii. −95 dBW/700 Hz in the 1610–1650 MHz frequency range.
> 3. Remediations
>   a. Ligado must expeditiously replace or repair as needed any U.S. government GPS devices that experience or are likely to experience harmful interference from Ligado's operations.[a]

b. Within 6 months of the release of FCC Order 20-48 or no less than 30 days prior to the deployment of downlink base stations at 1526–1536 MHz (whichever is sooner), Ligado is required to initiate an exchange of information with the U.S. government.
c. As part of this exchange program, Ligado must cooperate directly with any U.S. government agency that anticipates that its GPS devices may be affected by Ligado's ATC operations to:
   i. Provide base station location information and technical operating parameters to federal agencies prior to commencing operations in the 1526–1536 MHz band,
   ii. Work with the affected agency to identify the devices that could be affected and evaluate whether there would be harmful interference from Ligado's operation, and
   iii. Develop a program to repair or replace any such devices, said program consistent with that agency's programmatic needs, as well as applicable statutes and regulations relating to the ability of those agencies to accept this type of support.
d. Based on the base station and technical operating data made available to it, if an affected agency determines that Ligado's operations will cause harmful interference to a specific, identified GPS receiver operating on a military installation and that the GPS receiver is incapable of being fully tested or replaced, Ligado shall negotiate with the affected government agency to determine an acceptable received power level over the military installation.
e. Ligado must make available technical experts to support these repair and replacement programs.
f. Ligado must file quarterly reports with the Commission updating the Commission on the status and progress on Ligado's program to exchange information with federal agency GPS users on the deployment of base stations, identification of potentially affected devices, evaluation of risk of harmful interference, and development of a repair or replace program to meet federal agencies' needs.

4. Reporting Requirements
   a. At least 30 days before commencing transmission at a base station site, Ligado must submit to the FCC and the FAA a report that includes:
      i. the location of the proposed base station antenna site, and
      ii. the base station antenna radiation center height above ground level, its mechanical and electrical tilt, and its antenna specification, including polarization and pattern.
   b. Ligado must ensure this information is accessible to FAA stakeholders.

5. Ancillary Agreements
   a. Ligado must comply with the extant co-existence agreements in place with Garmin International, Inc., Trimble Navigation Limited, Deere & Company, NovAtel, Inc., Topcon Positioning Systems, Inc., and Hexagon Positioning Intelligence, as well as any additional such agreements into which it enters.

6. Continuation of ATC "Integrated Service" Rule Waiver
   a. The FCC reaffirmed its 2011 Order and Authorization which granted to Ligado predecessor SkyTerra a conditional waiver to the FCC's Integrated Service Rule which provides that an MSS operator must provide "substantial" satellite service and must offer "integrated" MSS and ATC service. The waiver is subject to the conditions that Ligado shall:
      i. Continue to make available and actively market a commercially competitive satellite service.
      ii. Continue to dedicate at least 6 MHz of MSS L-band spectrum, nationwide, exclusively to satellite service.
      iii. Continue to ensure that its satellite(s) are capable of operating across the entirety of Ligado's MSS L-band spectrum.
      iv. Continue to ensure that satellite-capable devices (both integrated MSS/ATC devices and satellite-only devices) using its MSS L-band spectrum are capable of operating across the entirety of Ligado's MSS L-band spectrum.

continued

> BOX 1-1 Continued
>
> v. Offer commercial satellite access agreements to terrestrial network operators on competitive pricing terms to enable integrated satellite and terrestrial service offerings for the Internet of Things (IoT).
> vi. Ensure that dual-mode MSS/ATC-capable L-Band IoT devices are available in the marketplace no later than September 30, 2024.
> vii. Undertake to standardize satellite IoT technology in the 3rd Generation Partnership Project to enable incorporation of satellite connectivity into chipsets consistent with its commitments to operate a satellite IoT network using such standards-based technology and to facilitate market participation through network access agreements for satellite IoT.
> viii. Ensure that it has the network capability to support MSS/ATC IoT devices and services using such standards-based technology.
> ix. Not offer preferential terms for the use of Ligado's spectrum for terrestrial only service, or otherwise discourage the availability or use of combined MSS/ATC services in addition to terrestrial-only services.
>
> 7. Ancillary Conditions
>   a. Ligado is subject to several reporting conditions related to its spectrum use roll out plans, any interference complaints and steps taken to remediate those complaints.
>
> ---
>
> [a] The committee interprets this to mean that Ligado would bear financial responsibility for these repairs and/or replacements, not that Ligado personnel would be the agents actually carrying out this work.

1.4 HARMFUL INTERFERENCE

Interference of a radio receiver occurs when some other signal affects the received signal in a way that reduces the effective received SNR. Functionally, an emitter causes interference whenever the operation of that device causes any noticeable reduction on the received SNR. There are various mechanisms through which an emitter could cause such a reduction. The most direct method is by emitting power directly into the receiver's signal band. Adjacent-band emitted power can also cause interference by a variety of mechanisms. Whether or not the interference is harmful to the receiver's operations depends on the amount of SNR loss and the analog and digital signal processing functions that the receiver must perform.

This section explores further how an out-of-band signal can interfere with a specific receiver. The discussion illuminates the mechanism of harmful interference and informs the analysis of impact of signal characteristics on existing GPS receivers. To ensure clarity, this section focuses on the broader meaning of "harmful interference" and not the FCC-defined term "Harmful Interference."

1.4.1 Received SNR and Interference Perceived as a Reduction in SNR

One way to understand the effects of SNR in GPS receivers and many other types of receivers is to look at the noisy symbols post correlation—that is, at the point in the signal processing where the Doppler-shifted carrier signal has been mixed out of the signal, any spreading code has been wiped off the signal, and the resulting signal samples have been accumulated. This discussion of SNR provides a cumulative view of any and all interference effects that might be caused by an adjacent-band signal as it enters the receiver's antenna and affects various stages of the receiver's RF front-end signal processing.

The analytic representation of the resulting noisy symbols are complex numbers of the form $I + jQ$, where $j = \sqrt{-1}$. Figures 1-7 and 1-8 plot typical Q versus I point clouds. These are also known as point constellations in the signal processing literature. Figure 1-7 is typical of a GPS receiver that processes a signal that carries navigation data that have been modulated onto the signal using binary-phase shift keying. That is, the digital data are either a 1 bit (phase angle = 0 deg), which centers its cloud on the $+I$ axis when the signal has been phase locked, or a 0 bit (phase angle = 180 deg), which centers its cloud on the $-I$ axis.

This committee has also been asked to consider the possible interfering effects of Ligado signals on MSS signals, such as Iridium. Figure 1-8 is typical of an Iridium handset processing a signal that carries communication data or navigation data that have been modulated using quadrature phase-shift keying (QPSK). This method encodes 2 bits per symbol. The four point clouds in the figure correspond to the four possible 2-bit combinations. The possible combinations are used to label their respective clouds in Figure 1-8. This figure corresponds to a case where phase-lock has not been achieved, hence the random angle by which the constellation is canted relative to the axes.

SNR depends on the distance of the center of each point cloud from the origin of the (I,Q) plane. This distance is denoted by the quantity A in the two figures. It also depends on the radius of each point cloud that captures 39.3 percent of the points in the respective cloud. This radius is denoted as σ in the two figures. The non-dimensional SNR equals $A^2/(2\sigma^2)$. The A^2 in the numerator represents the GPS signal's power. The σ^2 factor in the denominator arises from the noise in the I and the Q components, and the factor 2 arises because both noise components impact GPS receiver performance.[27] The noise is modeled as being independent and Gaussian in each component with a variance of σ^2. Absent the noise, each point cloud would collapse into a single point. Usually, the SNR is given in dB units, which is calculated using the expression $10log_{10}[A^2/(2\sigma^2)]$.

---

[27] Note that SNR can be split into a bit error SNR and a phase measurement error SNR rather than a single total SNR. In this alternative convention, these two SNRs both equal A^2/σ^2 rather than $A^2/(2\sigma^2)$.

A related quantity is the C/N_0. This equals the product of the SNR and the bandwidth, $BW = 1/T_{accum}$, where T_{accum} is the coherent integration time for the symbol(s) or code(s) being detected. Thus, $C/N_0 = A^2/(2\sigma^2 T_{accum})$. This quantity is nominally thought to give a good raw measure of the signal power relative to the average power spectral density of the received noise and to be independent of the signal processing within the receiver. As discussed in Section 1.2.4, this quantity is sometimes referred to as effective C/N_0. Often C/N_0 is given in decibel-Hertz (dB-Hz) units, which is calculated using the following formula: $10 log_{10}[A^2/(2\sigma^2 T_{accum})]$. The C/N_0, SNR, and T_{accum} values that correspond to the cases depicted in Figures 1-7 and 1-8 are given in their respective captions.

Interference, other than spoofing, manifests itself as a reduction of SNR, or equivalently, a reduction of effective C/N_0. The interference acts like added noise so that the receiver reports the value of $C/(N_0 + I)$ as though it were C/N_0, with I being the effective average power spectral density of the interfering signal as experienced by the receiver at the output of its RF front end. Interference causes a reduction of the ratio A/σ. That

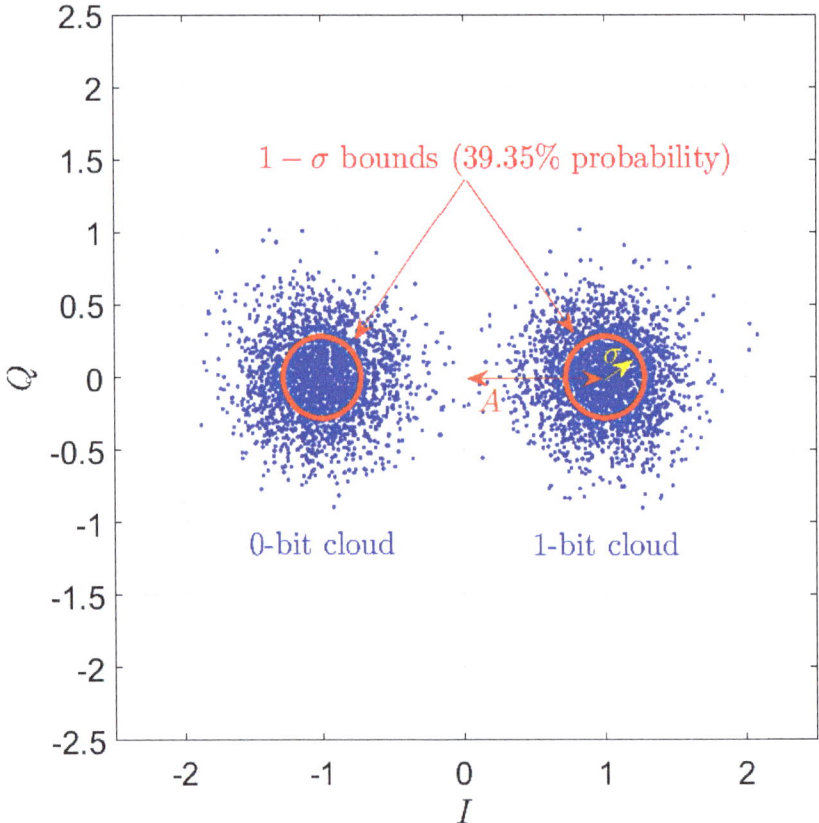

FIGURE 1-7 I/Q point clouds for a signal with BPSK data encoding at $\frac{C}{N_0}$ = 25 dB-Hz and an accumulation interval of T_{accum} = 0.020 sec, which yields SNR = 8 dB.

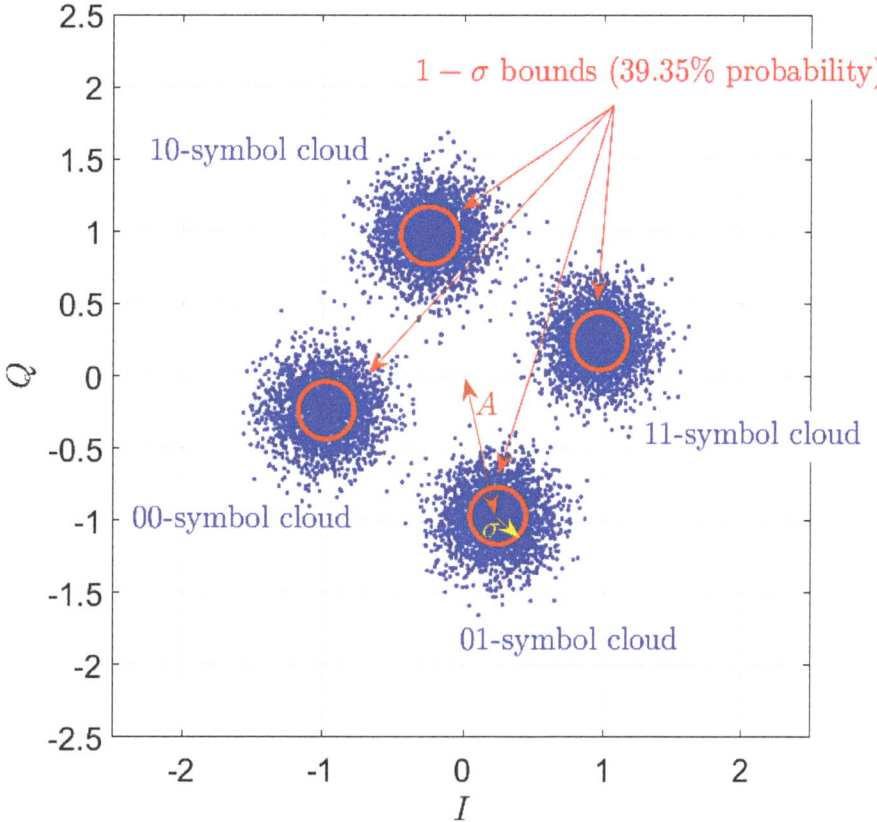

FIGURE 1-8 I/Q point clouds for a signal with QPSK data encoding at $\frac{C}{N_0}$ = 55 dB-Hz and an accumulation interval of T_{accum} = 0.000040 sec, which yields SNR = 11 dB.

is, the distance of the point cloud centers from the origin gets reduced relative to their diameters. This situation is depicted in Figure 1-9. The upper-left plot is a repeat of Figure 1-7 for a BPSK GPS signal, which assumes C/N_0 = 25 dB-Hz. The upper-right plot corresponds to a 1 dB interference-induced reduction of C/N_0, the lower-left plot corresponds to a 2.5 dB reduction, and the lower-right plot corresponds to a 4 dB reduction. If the interferer's power is strong enough or if it is near enough to the GPS receiver, then these levels of C/N_0 reduction or even more drastic levels of reduction can occur.

A similar situation occurs if the Iridium-type QPSK signal of Figure 1-8 experiences varying levels of effective C/N_0 reduction owing to interference. The four point clouds tend to merge as interference becomes more pronounced.

The effects of interference depicted in Figure 1-9 are shown as reductions in the signal power A^2 while the noise power $2\sigma^2$ remains constant. This is typically the case at the output of a GPS receiver's RF front end, where the effect typically manifests itself as an apparent reduction of A^2, consistent with the figure. This reduction occurs because a

Introduction and Background

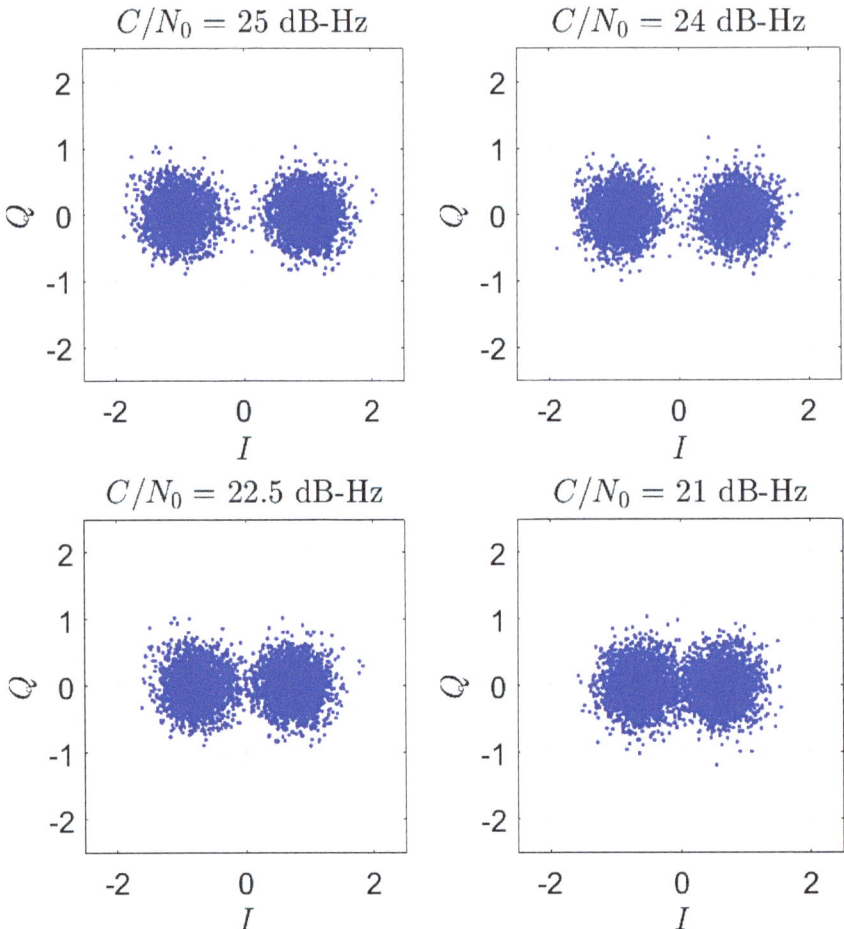

FIGURE 1-9 I/Q point clouds that illustrate the effects of progressively stronger interference for a signal with BPSK data encoding and an accumulation interval of $T_{accum} = 0.020$ sec.

typical receiver uses automatic gain control (AGC) in a situation where the noise power dominates the signal at the output of the RF front end owing to the spread spectrum nature of GPS signals. As noise-plus-interference power increases, the receiver's AGC decreases its gain in order to hold the power constant at the output of the RF front end's analog-to-digital converter (ADC). This AGC adjustment is needed in order to use the limited number of output bits of the ADC in an optimal manner under normal operating conditions.

Other receivers, such as Iridium receivers, may not use AGC. In such situations, the phenomenon of increased interference would manifest itself as an increase of the noise power $2\sigma^2$ while the signal power A^2 remained constant. In a figure such as Figure 1-8, this would correspond to a widening of the four point clouds while their centers remained a fixed distance from the origin. In either case, the point clouds tend to merge as interference increases.

1.4.2 The Problems That Interference Can Cause for a GPS Receiver

As noted in the FCC Order 20-48, the definition of "Harmful Interference" is:

> Interference which endangers the functioning of a radionavigation service or of other safety services or seriously degrades, obstructs, or repeatedly interrupts a radiocommunication service operating in accordance with [the ITU] Radio Regulations.

In order to assess the harmfulness of interference, it is helpful to understand the physical mechanisms by which harm occurs. The progressive effects of increasing interference, as depicted in Figure 1-9, can cause problems for GPS receivers in a number of ways. The first question put to this committee asked whether a given level of SNR reduction or a given level of position accuracy degradation was the best indicator of harmful interference. This reduction of the question of harmful interference to a simple binary choice obscures important aspects of effective GPS receiver operation. There are multiple ways in which interference can degrade GPS receiver operation that defy simple characterization as either a single SNR degradation metric or a position accuracy degradation metric.

Although much of the current discussion is framed in terms of the effects of the impacts of SNR reductions owing to interference, this approach is used as a means of illustrating the harmful effects of interference from a GPS receiver's viewpoint. Considerations from this viewpoint, however, do not allow the prediction of harmful interference by a single value for a tolerable amount of SNR loss or, equivalently, C/N_0 loss.

One problem that can occur for a GPS receiver is an inability to correctly decode the navigation data that are broadcast by the GPS satellites on the signals and that are needed by the receiver in order to compute GPS satellite locations and clock calibration offsets, which are necessary inputs to a navigation solution. As C/N_0 decreases and the pairs of point clouds in Figure 1-9 begin to merge, bit error rates increase, and eventually the receiver will be unable to decode the needed data. At this point, the given GPS signal becomes unusable for navigation purposes.

Another problem that is caused by interference concerns degradation of a receiver's ability to measure the amplitude and phase of the (I,Q)-plane point clouds, as depicted in Figure 1-10. As the effective C/N_0 decreases owing to interference, the signal amplitudes A_i and beat carrier phase offsets $\Delta\phi_i$ of individual points within a cloud become less and less accurate measures of the amplitude and phase offset of the corresponding cloud center. (Note that the term "beat carrier phase" refers to the difference between the nominal phase that the received signal would have if there were no carrier

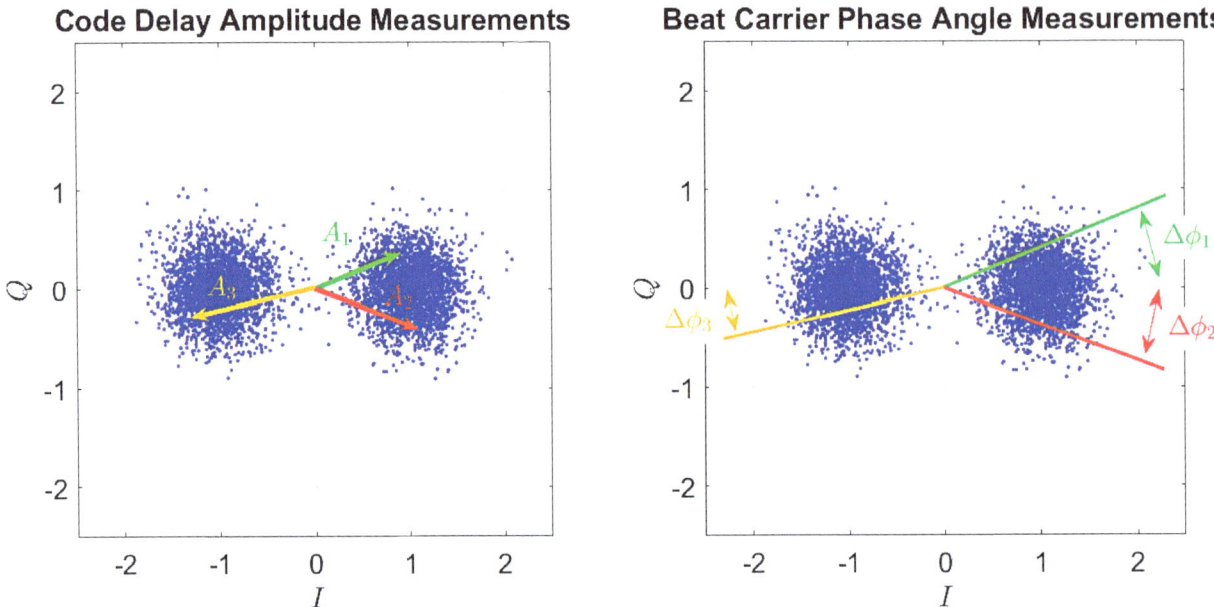

FIGURE 1-10 Examples of the (I,Q)-plane amplitude measurements (left panel) and phase measurements (right panel) that provide the bases for GPS signal tracking and for the inference of navigation observables.

Doppler shift and the actual phase that it has. Thus, the beat carrier phase is the negative of the time integral of the carrier Doppler shift.)

Errors in individual A_i values as measurements of a cloud center's distance from the origin of the (I,Q) plane translate indirectly into pseudo-random number (PRN) spreading code phase measurement errors by the receiver. The measurement precision of PRN spreading code phase translates directly into pseudo-range measurement precision, which is the fundamental measurement of standard GPS position and time offset determination. Increases in the root mean square (RMS) noise-plus-interference component of a receiver's navigation error are proportional to increases in the RMS value of the A_i measurement errors expressed as a fraction of the true received value. Thus, the RMS value of the positioning error component is proportional to $1/\sqrt{C/N_0}$. Furthermore, if the RMS code phase measurement errors become too large, then the receiver can lose code lock on the signal, in which case it becomes unusable for surveying, scientific applications, and navigation.

Errors in individual $\Delta\phi_i$ values as measurements of a cloud center's rotational orientation relative to the I axis directly affect carrier Doppler shift measurement errors if the receiver uses a frequency-lock loop (FLL). If it uses a phase-lock loop (PLL), then $\Delta\phi_i$ measurement errors translate directly into beat carrier phase errors in this loop's feedback path, and they also affect its estimation of carrier Doppler shift. The measurement precision of carrier Doppler shift translates directly into range-rate measurement precision, which is the fundamental measurement of standard GPS velocity and time offset rate determination. The measurement precision of the beat carrier phase translates directly into accumulated delta range measurement precision, which is the fundamental observable of high-performance (HP) GPS ultra-precise position and time offset determination. Increases in the RMS noise-plus-interference component of a receiver's velocity error or its high-precision position error are proportional to increases in the RMS value of the $\Delta \phi_i$ measurement errors expressed in radians. Thus, the RMS value of the velocity error component or its high-precision position error component is proportional to $1/\sqrt{C/N_0}$. Furthermore, if the RMS frequency or beat carrier phase measurement errors become too large, then the receiver can lose frequency lock or phase lock on the signal, in which case it becomes unusable for navigation.

Even before loss of phase lock, a PLL can start to experience cycle slips. As C/N_0 decreases in a given HP receiver, the frequency of cycle slipping increases. These slips are highly detrimental to the performance of an HP receiver because the receiver cannot use the given signal in its ultra-precise position fixes until the integer number of cycle slips has been resolved. This can take minutes. This can cause a delay of minutes to hours in a HP application such as surveying. For the HP application of precise construction site grading using automated earth-moving equipment, the entire operation could be placed on hold because such equipment needs precise position fixes in real time. For the HP application of tsunami prediction, the system would fail to provide any usable data.

1.4.3 The Variability of Tolerable Amounts of C/N_0 Loss

The amount of loss of C/N_0 that is tolerable depends on the signal and on the receiver's signal processing architecture. Consider the situation of a clear sky view, negligible adjacent-band interference, and a high-sensitivity receiver that uses a good patch antenna with an embedded LNA. Received C/N_0 values for the visible GPS satellites might range from 50 dB-Hz for high-elevation satellites down to 35 dB-Hz for satellites near the horizon. A signal with C/N_0 = 50 dB-Hz could easily tolerate an interference-induced degradation of 5 dB. A signal with C/N_0 = 35 dB-Hz might be rendered useless by a 5 dB degradation. Thus, the tolerable degradation depends on the likely range of available C/N_0 values for the nominal application in the absence of interference. Furthermore, environmental and use factors may consume this link margin—for example, building or foliage obstruction, ionospheric scintillation, or vibration and dynamics. A typical receiver design includes link margin for such situations. It might not have enough margin to handle these design cases along with degradation from Ligado, even if the latter were less than 5 dB.

Furthermore, there can be a lot of variability between receivers of their minimum usable C/N_0 values. Recalling Figure 1-10, an obvious thing to do in order to improve both the A_i measurement accuracy and the $\Delta\phi_i$ measurement accuracy would be to average the locations of multiple (I,Q) points that were known to lie in the same cloud. If C/N_0 gets degraded too much, then the point clouds start to coalesce, and this method can break down. For Assisted GPS (A-GPS), as in the E911 services of smartphones, the point cloud memberships can be known a priori via cellular-network-supplied data, and there is no such problem of proper point/cloud association. A-GPS receivers can operate at very low C/N_0 values. Other receivers have hard lower limits of about 20–25 dB-Hz for proper point-cloud association for the GPS L1 C/A codes that form the foundation of much of the currently used civilian GPS.

For various reasons, however, C/N_0 in the range 20–25 dB-Hz may not be tolerable for some receivers. They may need to track dynamic signal variations with a high bandwidth. This precludes the use of too much averaging of (I,Q) points before the extraction of A_i and $\Delta\phi_i$ measurements. Some receivers cannot use signals with C/N_0 values lower than 35 dB-Hz.

Another factor in receiver performance is the number of usable GPS signals. The bare minimum for standard navigation is four. Typical receivers see eight or more signals and use all of them in order to get an improved solution. Interference that caused a 5 dB loss of C/N_0 might knock out the four weakest signals but only degrade the accuracy of the other four signals by tolerable amounts. If the geometry of the remaining four satellites is good, then the achieved RMS position error might not be degraded significantly. If the receiver also needed to perform Receiver Autonomous Integrity Monitoring (RAIM), that capability would be lost owing to the lost "extra" satellites. HP applications require far more than the minimum of four satellites in order to achieve mm-precisions and to resolve carrier cycle slips.

Yet another problem with degraded C/N_0 is that of signal acquisition, especially cold-start acquisition. Cold-start acquisition involves a brute-force search for signal power above the noise floor in a two-dimensional space of possible carrier Doppler shifts and PRN code delays. For low values of C/N_0, the required amount of coherent integration time, T_{accum}, increases markedly, and this increase forces the use of a finer grid in the Doppler shift search space. The search will eventually take too long or become impossible owing to point/cloud ambiguities. As an example, a given receiver may be able to maintain lock and deduce navigation observables from a signal with a C/N_0 value of only 25 dB-Hz, but it might be unable to perform successful cold-start acquisition for a signal with C/N_0 less than 35 dB-Hz. Such a problem might not be evidenced in a controlled test of RMS position error for receivers that experienced interference after they had acquired the GPS signals and successfully transitioned to tracking.

1.4.4 Mechanisms by Which an Adjacent-Band Emitter Causes SNR (and C/N_0) to Degrade and the Variability of These Mechanisms Between Receivers

Interference from an adjacent-band emitter manifests itself in a receiver as a noticeable reduction in A/σ when the emitter is turned on or when the receiver comes close enough to the emitter. That is, the point clouds in Figure 1-9 progress from left to right and from top to bottom for a given GPS signal as processed within a given receiver.

There are five mechanisms by which an adjacent-band emitter can cause a reduction in SNR or, equivalently, C/N_0. The following discussion of these mechanisms borrows heavily from a paper by Hegarty et al. (2021).[28] Four of the five mechanisms are functions of the receiver's RF front-end design. The second author of the cited paper is an internationally recognized expert in the field of GPS receiver RF front-end design, who has designed front ends for commercial GPS receiver manufacturers and for specialty research applications, among them the FOTON receiver that currently flies on the International Space Station and that has been used to detect a GPS jammer at a Russian base in Syria.[29]

The most obvious mechanism for impacting C/N_0 is the OOBE of the emitter that fall into the GPS band. These emissions contribute directly to the effective average value of N_0 in the ratio C/N_0. GPS receivers cannot mitigate this component of interference unless it has special properties, such as lying in a narrow frequency band or in a narrow time window. If it is too large and does not have such special properties, then all GPS receivers' performance will degrade significantly.

The other four mechanisms for an adjacent-band emitter to cause C/N_0-reducing interference all involve the GPS receiver allowing adjacent-band signals to enter the GPS receiver pass band directly or indirectly, or they involve the receiver reducing the in-band power of the GPS signal as a by-product of the effect of the out-of-band signal. Two of these mechanisms involve nonlinear effects in the receiver's RF front end. These nonlinearities exist in the receiver's LNAs, its mixers, and its other circuitry. Adjacent-band power, if it enters an LNA, can drive the LNA into its nonlinear range where the effective gain gets reduced. This is referred to as compression. This gain reduction affects the in-band GPS signal, thereby reducing the value of the received signal power C in the ratio

---

[28] C.J. Hegarty, D. Bobyn, J. Grabowski, and A.J. Van Dierendonck, 2020, "An Overview of the Effects of Out-of-Band Interference on GNSS Receivers," *NAVIGATION* 67(1):143–161.

[29] M.J. Murrian, L. Narula, P.A. Iannucci, et al., 2021, "First Results from Three Years of GNSS Interference Monitoring from Low Earth Orbit," *NAVIGATION* 68(4):673–685.

C/N_0. Another nonlinear effect is intermodulation. The adjacent-band power may inadvertently get mixed with a signal in another adjacent band owing to receiver nonlinearities, and this spurious mixing may send power into the GPS band, effectively increasing the noise in the GPS band.

The effect known as reciprocal mixing occurs if the mixing signals used by the receiver are not perfect sinusoids. The receiver multiplies these signals by the incoming RF signal and later by intermediate-frequency (IF) versions of the original signal in order to convert it to a lower IF or even to base-band. The output frequency is the difference of the original signal's RF or IF and the mixer frequency—if the latter is a pure sinusoid. If the mixer has any harmonics, then they could mix with adjacent-band signals in a way that translates those signals into the new IF of the GPS signals. If this happens, then the adjacent-band power gets overlaid on top of the GPS power.

The last mechanism by which adjacent-band power gets translated into the GPS band is simple aliasing. If f_s is the sample rate, then the Nyquist band extends from $-f_s/2$ to $+f_s/2$. Any remaining signal components outside this band prior to digital sampling will be shifted into this band via a frequency shift equal to mf_s, where m takes on whatever integer value is needed in order to shift the component into this band. Attenuated, adjacent-band power after the final analog processing on the RF front end can get aliased into a digital GPS receiver's Nyquist band by the sampling of the RF front end's ADC at some sampling rates. Depending on the receiver, the Nyquist band is typically not much wider than the GPS signal of interest. Under these circumstances, the aliased energy is likely to interfere with the GPS signals.

LNA compression and intermodulation, being nonlinear effects, likely will not exhibit a one-to-one relationship to the adjacent-band signal power. Thus, a 1 dB increase in the received adjacent-band power will likely not translate into a 1 dB decrease in C/N_0. The decrease may be more or it may be less than 1 dB. This type of loss relationship is illustrated in Figure 1-11, which shows the results from a test that have been supplied by the DOT. The value of C/N_0 for received interference powers below −50 dBm does not change. This happens because the effective added noise power spectral density remains negligible relative to the receiver's nominal N_0 owing to thermal noise. Above −40 dBm the C/N_0 versus interference power approaches a negatively sloped asymptote, with slope on the order of −10 dB/2 dB = −5 dB/dB. Although it is possible to conclude from this figure that a nonlinear effect such as LNA compression, intermodulation, or some combination of these effects is present for the tested receiver model, it is not possible to make a determination among these causes without detailed information about the particular receiver design's RF front end.

Reciprocal mixing and aliasing, on the other hand, are linear on the adjacent-band signal. Therefore, one would expect the high-interference-power asymptotes of their C/N_0 versus noise power curves to have slopes of −1 dB/dB. This is nearly the high-interference-power asymptotic slope for the identical pair of receivers whose DOT-provided test results are shown in Figure 1-12. As with the data in Figure 1-11, it is not possible to know whether this receiver model's C/N_0 degradation results from reciprocal mixing, aliasing, or some combination thereof without detailed information about its RF front end's design.

1.4.5 Receiver Design Impact on the Ability of Adjacent-Band Power to Impact the Effective C/N_0

The four mechanisms by which adjacent-band power impacts GPS C/N_0 can be reduced by the designs of the RF front-end electronics. In particular, various stages of band-pass filtering are applied in a typical RF front end. If designed properly, the impact of adjacent-band power can be reduced by many 10s of dB. A good RF front-end designer considers the anticipated adjacent-band power levels and designs the RF front-end LNA,

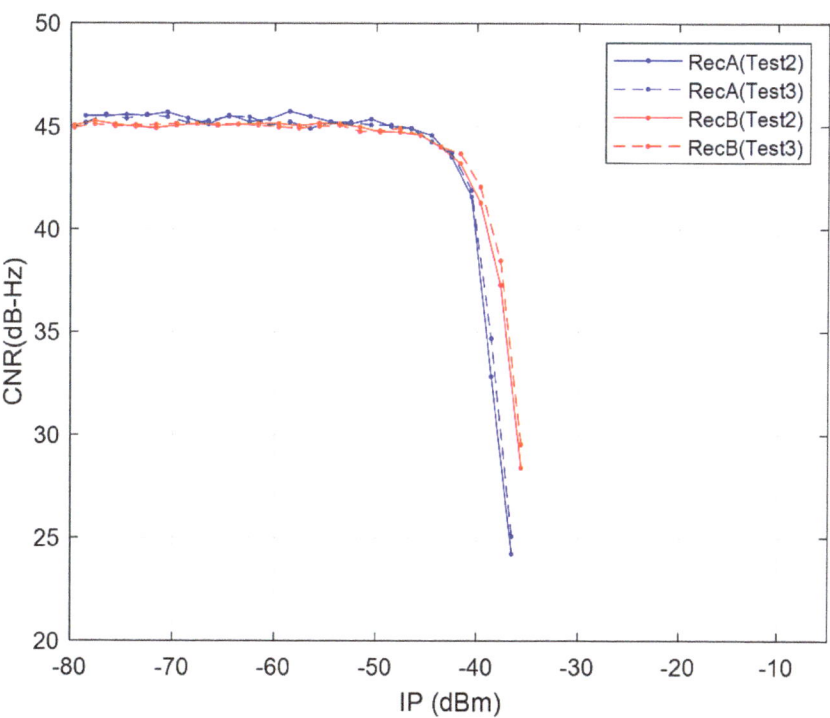

FIGURE 1-11 C/N_0 versus received power from a Ligado tower power for two identical receivers and two tests.
SOURCE: U.S. Department of Transportation Volpe Center, 2022, "Response to National Academies of Sciences, Engineering, and Medicine Questions," Document submitted to the Committee to Review FCC Order 20-48, Washington, DC: National Academies of Sciences, Engineering, and Medicine.

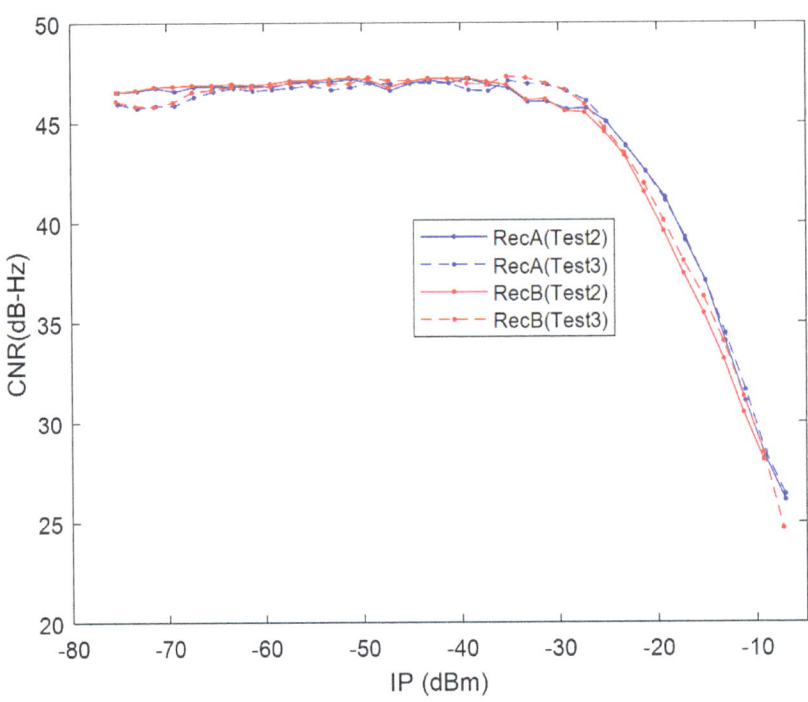

FIGURE 1-12 C/N_0 versus received power from a Ligado tower power for a different pair of identical receivers from those of Figure 1-11, which also undergo two tests.
SOURCE: U.S. Department of Transportation Volpe Center, 2022, "Response to National Academies of Sciences, Engineering, and Medicine Questions," Document submitted to the Committee to Review FCC Order 20-48, Washington, DC: National Academies of Sciences, Engineering, and Medicine.

mixing, and filtering scheme appropriately in order for the net impact of all adjacent-band power to be negligible. If the adjacent-band spectrum undergoes changes, as is the case with FCC Order 20-48, an RF front-end design that worked very well pre-FCC Order 20-48 may or may not work well post-FCC Order 20-48.

Figure 1-13 plots the adjacent-band power levels that cause a 1 dB C/N_0 reduction as functions of the adjacent-band power's center frequency. The Ligado tower frequency band center is 1531 MHz. The various curves apply to a variety of tested GPS receivers. This plot has been supplied to the committee by the DOT and shows that the power levels required to produce a 1 dB C/N_0 reduction in the tested receivers varies from a high of −15 dBm to a low of −75 dBm. Thus, there is a 60 dB variation of receiver susceptibility to interference from Ligado towers as measured using a 1 dB C/N_0 loss threshold. This wide variation of sensed interference at each receiver is not owing to Ligado's out-of-band emissions from the tower that fall within the GPS band, because that mechanism of interference is virtually identical for all GPS receivers. The variations must arise owing to one or the other of the four additional mechanisms by which adjacent-band power can impact GPS in-band C/N_0, all of which are highly dependent on the receiver front-end filtering: LNA compression, intermodulation, reciprocal mixing, or aliasing. Which

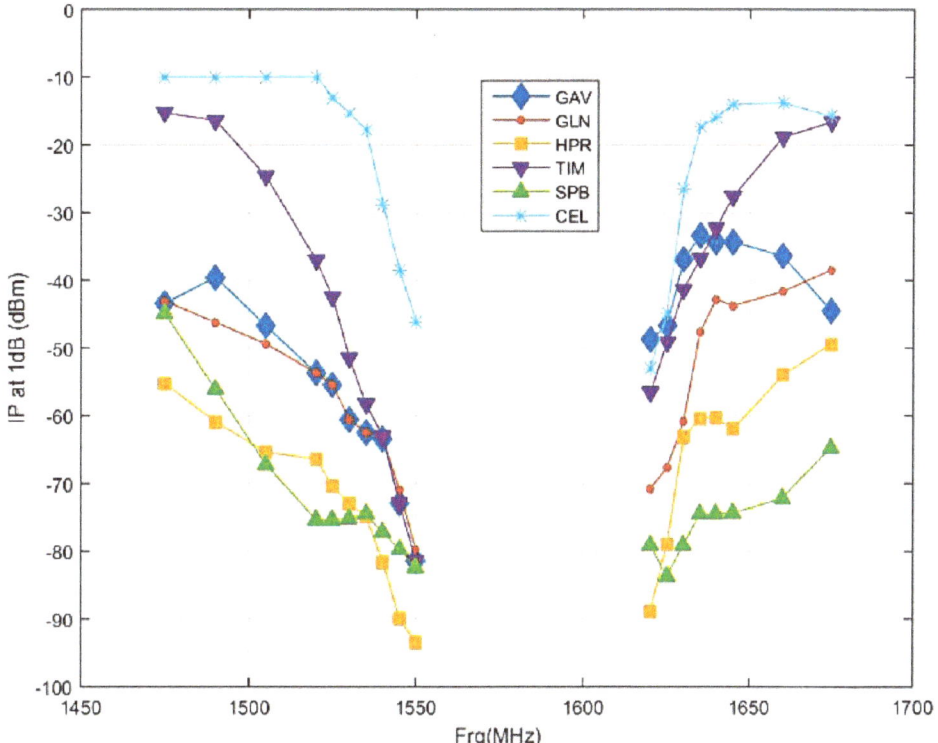

FIGURE 1-13 Adjacent-band power levels that cause a 1 dB reduction in C/N_0 as reported by the receiver versus the center frequency of the adjacent band. The types of receivers shown are General Aviation (Non-Certified) (GAV), General Location/Navigation (GLN), High Precision (HPR), Timing (TIM), Space Based (SPB), and Cellular (CEL).
SOURCE: U.S. Department of Transportation, 2018, *United States Department of Transportation Global Positioning System (GPS) Adjacent Band Compatibility Assessment*, Washington, DC, Figure 3-22.

mechanism or combination of mechanisms applies to a given receiver cannot be determined from the plot. What is clear is that the best receiver does a much better job relative to the worst receiver of attenuating the Ligado signal before it can affect the GPS in-band C/N_0.

1.4.6 Summary Regarding Harmful Interference

Harmful interference of a given GPS receiver owing to an adjacent-band signal occurs when that signal causes the effective C/N_0 values of enough signals to degrade by large enough values to degrade the receiver's performance. The amount by which C/N_0 actually gets degraded is highly dependent on receiver design.

Thus, any standard purporting to predict harmful interference must first characterize the power of a potentially interfering signal in specific frequency bands. Limiting interfering power in the RNSS band (1559–1610 MHz with the GPS L1 signals occupying part of this band) is of primary importance to preventing harmful interference; however, interfering power in adjacent frequency bands can also cause harmful interference. The effect of interference power in adjacent frequency bands diminishes as the frequency of the interfering signal gets farther and farther from the intended reception band. Harmful interference as it affects a receiver is dependent on receiver design; so, any standard that purports to predict harmful interference must make assumptions about "reasonable" receiver designs. Indeed, "certified receivers" do just that. Furthermore, C/N_0 degradation can be caused either by increased interference power, decreased signal power, or receiver design; so, it is not a workable regulatory standard for specifying emitter characteristics. However, the foregoing discussion helps us understand the mechanisms of harmful interference on receivers and can lead to a reasonable regulatory standard that specifies tolerable (and intolerable) interference power in specific frequency ranges that include bands that are adjacent to the RNSS band and not just the actual RNSS band.

It is possible to design L1 GPS RF front ends so that the degradation of the effective C/N_0 is negligible for the permitted Ligado signal power characteristics even at close ranges. Not all existing GPS receivers have RF front ends that do this, although almost all receivers tested before the Ligado order experienced no harmful interference. In Task 2, the committee examines the likelihood of harmful interference owing to operation of Ligado emitters as specified in the FCC order.

Even for a given level of C/N_0 degradation, the question of whether it causes harm is receiver-dependent and application-dependent. For some low-bandwidth applications, a large reduction in C/N_0 may be tolerated by the receiver without any significant loss of function. For other receivers, a C/N_0 reduction of 4 dB or even less could cause significant harm to performance. This harm could come in a variety of forms. It could involve reduced position accuracy. It could involve increased frequency of cycle slips that degrades the availability of high-precision position solutions or altogether precludes them. It could involve an inability to acquire new signals. It could involve an inability to perform Receiver Autonomous Integrity Monitoring. There is no way to know ahead of time what level of C/N_0 degradation may cause harm and how that harm may manifest itself unless one analyzes a particular receiver's ability to support a particular application.

Even a receiver that can tolerate a large degradation of C/N_0 in open-sky conditions may not tolerate a much smaller loss of C/N_0 in other situations for which it has been designed. Consider a smartphone receiver that supports Enhanced 911 (E911) service. It is designed to work with very low C/N_0 values in order to continue providing location to emergency responders while deep inside a building. That capability would help make it insensitive to the permitted Ligado signals when operating in an open-sky test or the equivalent. Suppose it were operating inside a building that included an indoor Ligado "tower." The combined reduction in C/N_0 owing to being indoors and owing to the Ligado signal could render its E911 geolocation capability inoperable even though it could have supported E911 service if it only had to contend with being indoors, as per design.

Test results from one receiver manufacturer indicate that their newest high-precision receivers can operate satisfactorily in the presence of the currently approved Ligado downlink power level. Therefore, it is known that the technology currently exists to design GPS receivers for some functions, perhaps some of the most demanding functions, that will not experience harmful L1-band interference from Ligado downlink signals. This has been achieved through proper design of the RF front end to attenuate the Ligado signals sufficiently before they can negatively impact the effective in-band C/N_0 to a noticeable extent.

Some receivers, both commercial and DoD, process signals from multiple bands—for example, L1 and L2, or L1, L2, L5, Galileo, BeiDou, and GLONASS. They may use a Kalman filter in order to fuse the various pieces of radionavigation data from the various bands and systems. In some cases, the data from other bands might mitigate harmful interference to data in the L1 band owing to Ligado signals. In other cases, the use of multiple bands may just shift the mode by which Ligado interference on the L1 band harms receiver performance. For example, suppose that an L1/L2 receiver has been configured to use the data from its two frequencies in order to estimate and cancel the ionosphere delay. In that case, interference on L1 owing to Ligado could disable this receiver function so that harm would come in the form of increased receiver navigation error owing to increased ionosphere error. Questions of how Ligado signals might or might not harm multiband receivers have not been deemed to be in detail by the committee. Such questions were considered beyond the scope of its inquiry. Furthermore, receiver testing did not consider such issues.

1.4.7 Definition of Harmful Interference

The FCC included the following definition of Harmful Interference in FCC Order 20-48:

> Interference which endangers the functioning of a radionavigation service or of other safety services or seriously degrades, obstructs, or repeatedly interrupts a radiocommunication service operating in accordance with [the ITU] Radio Regulations.

In addition, the FCC uses the following very similar definition from United States Code (USC), Title 47 Telecommunications, Chapter 1, Subchapter A, Part 15, Section 15.3 Definitions:

(m) Harmful interference. Any emission, radiation or induction that endangers the functioning of a radio navigation service or of other safety services or seriously degrades, obstructs or repeatedly interrupts a radiocommunications service operating in accordance with this chapter.

These paragraphs define Harmful Interference for two classes of radiocommunication services: (1) radionavigation and other safety services and (2) all other radiocommunication services.

Two conditions must—per provisions of USC, Title 47 Telecommunications, Chapter 1, FCC, Within the United States and Possessions—be met in each case for a situation to be considered Harmful Interference:

1. The emission, radiation or induction must (a) endanger the functioning of the radionavigation or safety service or (b) seriously degrade, obstruct, or repeatedly interrupt any other radiocommunication service; and
2. The service claiming Harmful Interference must be operating in accordance with regulations.

The definition of Harmful Interference is effectively the same for the ITU, which regulates international frequency spectrum usage, the NTIA, which manages U.S. federal government use of spectrum, and the FCC, which regulates U.S. commercial as well as state and local government use of spectrum.

Given that there is greater likelihood of loss or life or property when a radionavigation or safety service is interfered with, the threshold conditions for Harmful Interference are lower than those for other radiocommunication services.

GPS operates in the RNSS radiocommunication service, so Harmful Interference, as specified by the FCC, is defined by items 1a and 2. Iridium operates in the MSS, so Harmful Interference is defined by items 1b and 2.

By design, "Harmful Interference" is defined, a priori, in subjective operational terms, rather than technical, objective terms. Also, some interference does not rise to the level of being harmful, as discussed already in this section. The FCC initially evaluates Harmful Interference on a case-by-case basis. This is to allow regulators to balance the relative benefits of competing systems and technologies. In many, but not all instances, once competing and complementary interests are reconciled, technical rules and regulations can be written to define the boundaries of "Harmful Interference" as purportedly accomplished in FCC Order 20-48.

This committee was charged with reviewing whether FCC Order 20-48 actually creates conditions such that no harmful interference will occur as a result of its permitted Ligado emissions. The committee has concentrated on the physics and engineering questions of harmful interference, but not on the strictly legal definitions. Therefore, the committee made no effort to assess whether any RNSS or MSS—for example, GPS and Iridium—were or were not "operating in accordance with [the ITU] Radio Regulations" or "operating in accordance with this chapter." Instead, the committee proceeded under the assumption that it has been charged with determining whether existing RNSS or MSS would be harmed by Ligado interference independent of any legal ruling about whether they were "operating in accordance with [the ITU] Radio Regulations" or "operating in accordance with this chapter."

2

Analysis Regarding the Three Study Tasks

2.1 TASK 1: APPROACHES TO EVALUATING HARMFUL INTERFERENCE CONCERNS

The following task is the first that has been put to this committee to consider:

Which of the two prevailing proposed approaches to evaluating harmful interference concerns—one based on a signal-to-noise interference protection criterion and the other based on a device-by-device measurement of the Global Positioning System (GPS) position error—most effectively mitigates risks of harmful interference with GPS services and U.S. Department of Defense (DoD) operations and activities.

In this task, the committee is charged with resolving a dispute between the GPS community and Ligado about which of the two methods is best for characterizing harmful interference. The committee's response, discussed in this section, is that neither method is satisfactory in the simplistic forms in which they have been proposed.

2.1.1 Overview of Response to Task 1

As they were applied, neither of the prevailing approaches effectively mitigates the risk of harmful interference. As implemented, the signal-to-noise ratio (SNR) approach was too inflexible, and the position approach was too narrow in its applicability. Yet, both approaches do have a role in evaluating harmful interference to existing receivers. Of the two, the SNR-based approach is the more comprehensive and informative. By informing the degradation in link margin,[1] this approach can be used to predict harmful impacts across a broad set of use cases: signal acquisition, real-time kinematic (RTK), timing, and so on. Neither approach provides an analytical, repeatable, or straightforward criteria to evaluate new entrants.

The determination of harmful interference is dependent on the particulars. There is an incredibly wide array of GPS use cases, including car navigation, network timing, precise surveying, and geophysical applications, to name only a few. These use cases have different failure modes, which result in varying interference tolerance. The most appropriate approach must be mapped to each use case. As such, the use of a single, firm SNR-based interference protection criterion (IPC) is not practical when applied to device-by-device performance. Although evaluation of device-by-device SNR degradation is informative and recommended, the criteria must incorporate the fact that harmful interference depends on use case. For example, some applications are harmed when code-lock is lost, while other applications are harmed from loss of carrier phase-lock.

The commonly advocated 1 dB SNR loss criteria has not been linked to the definition of Harmful Interference. Although a "1 dB criterion" precludes harmful interference in virtually all use cases, the vast majority of GPS use cases do not experience harmful interference at such a low level—that is, they have significantly higher link margin than 1 dB. As such, the 1 dB criterion is prophylactic, but conservative.

Task 1, as posed, does not directly address the bigger challenge: regardless of which approach is applied, drastically different conclusions can be reached. There are numerous test design particulars that must be considered, including determining the path-loss model, desired stand-off range, antenna coupling, degree of insensitivity of a particular receiver's design to adjacent-band power, and performance thresholds. Even for a given use case, these issues are not easily resolved. Furthermore, a per-device SNR threshold creates a moral hazard—receiver manufacturers are incentivized to design adjacent-band-susceptible receivers in order to claim spectral easements.

Ultimately, both of the proposed approaches are cumbersome, owing to the intensive testing required. They do not provide an engineerable, predictable standard that new entrants can readily use to evaluate impact. As such, these approaches impede spectral progress. A new applicant for emissions in a new adjacent channel will have great difficulty in determining the emitter power levels and stand-off distances that will be guaranteed not to cause harmful interference to the existing GPS receiver base. A GPS receiver designer will be unable to design a receiver that will be guaranteed to be insensitive to unknown potential future allowed levels of adjacent-band power that might be allowed if the Federal Communications Commission (FCC) were to amend its Order

---

[1] "Link margin" refers to the difference between the observed post correlation SNR (in dB) and the minimum operating threshold (in dB).

20-48 to allow additional adjacent-band power beyond that contained in Box 1-1 in this report.

The next two sections discuss these two approaches in more detail. Their strengths and weaknesses are noted.

2.1.2 Considerations Regarding an SNR IPC

The SNR IPC approach consists of testing GPS receivers at various interference levels that could be attributed to Ligado transmissions and measuring the resulting degradation in reported C/N_0. C/N_0 is the standard industry-wide metric that is accepted by GPS subject matter experts to measure the operational health of a GPS receiver. Many key receiver functions can be mapped to C/N_0 thresholds—that is, C/N_0 level can be used to reliably predict a receiver's ability to acquire signals in cold-, warm-, and hot-start. It can also be used to predict code- and carrier-lock. As such, it informs receiver performance over a wide set of operating conditions, including RTK. Insight into these core functions can be extrapolated to four GPS performance metrics: accuracy, continuity, integrity, and reliability.

To be clear, the measurand is the *receiver reported* C/N_0. This is computed internally by the GPS receiver, within its digital logic, typically measured from the correlator outputs, as described above. This is a device-estimated quantity, which is noteworthy because Task 1 implies that the C/N_0 criteria is a physical standard independent of receiver design and thus subject to purely analytical assessment, which is incorrect. Furthermore, because this measurement is made after the correlation, it blends the effects of both in-band and adjacent-band interference, with the latter being dependent on the receiver's filtering design. In many regards, the intermingling is ideal because the receiver-reported C/N_0 lumps together all impacts from the interference. Nevertheless, it should be observed that there is not a deterministic mapping between receiver-reported degradation of effective C/N_0 and the physical power of the radio frequency (RF) wavefront as it arrives at the antenna, especially the adjacent-band power.

Generally, receiver-reported C/N_0 can reliably and consistently measure changes to better than 1 dB of accuracy. The key here is that the *change* in C/N_0 is being measured. There will be biases in the absolute C/N_0 estimate. That bias, however, falls out when looking at changes. The claim of "better than 1 dB accuracy" is a consequence of two effects: (1) the standard C/N_0 estimators are sufficiently linear over the region of interest (20–50 dB-Hz), and (2) the noise in the estimators can be effectively averaged down in reasonable bandwidths (averaging with less than a minute of data is sufficient). The accuracy of the delta C/N_0 measurement is only for a specific device. Across devices, there will be variation owing to implementation specifics (hardware and firmware), as previously discussed. There is a cause for caution because, as an internally generated figure, there is no guarantee that every device's receiver reported C/N_0 is computed using a widely accepted calculation. Additionally, real-world, live sky testing needs to be considered carefully because the carrier power can vary over time from multiple causes—satellite motion, antenna tilt, multipath, and so on. Simulator testing with controlled, or constant, signal power removes this concern.

As discussed earlier in this chapter and in the explanation of interference in Section 1.4, C/N_0 degradation is highly predictive of receiver performance, although identifying the cause of the degradation is less predictive and is receiver dependent. It can be used to inform impact across a range of regimes: cold-start, warm-start, code-tracking, and carrier-tracking. The required C/N_0 thresholds for these operational modes are known to provide reasonable accuracy for most receivers. As such, receiver-reported C/N_0 degradation provides insight into how a given GPS receiver's performance will be impacted across a variety of use cases. Harmful interference must be tied to the available margin a receiver has for its intended task. Some applications, such as a car-based GPS, may depend only on code-lock. RTK and scientific applications typically depend on phase-lock and on mean time between cycle slips. A stationary receiver mounted adjacent to a Ligado transmitter may be limited by acquisition performance. The thresholds for these applications can vary significantly, by tens of decibels. Therefore, the same receiver model may be impacted in one use case but not another. As such, a single hard IPC threshold for C/N_0 is inadequate.

In practice, the question of how much degradation occurs to C/N_0 is highly receiver dependent. The Ligado emissions that fall directly into the GPS band can be considered separately from the emissions that fall outside the GPS band. The impact from Ligado emissions in the GPS band is not receiver dependent. This portion raises the noise floor and can be computed analytically. The Ligado signal components outside the GPS band affect C/N_0 in the GPS band via the GPS receiver RF front-end electronics (see Section 1.4.5). An important takeaway is that, depending on the receiver design, the levels of power that get translated from adjacent-band to GPS in-band can vary by 60 dB or more, as evidenced by Figure 2-1. In the Ligado downlink band from 1526–1536 MHz, there is a 60–65 dB variation of the emitted out-of-band power levels that result in a 1 dB reported loss of C/N_0 for various classes of receiver. It is reasonable to infer that these large differences in reported C/N_0 degradation stem from large differences in the various receiver classes' ability to tolerate adjacent-band signal power. All of these receivers reject adjacent-band power to a sufficient degree to operate well in the pre-Ligado adjacent-band spectrum for which they were designed. Some, however, appear to have insufficient adjacent-band power attenuation once the approved Ligado downlink signals are present.

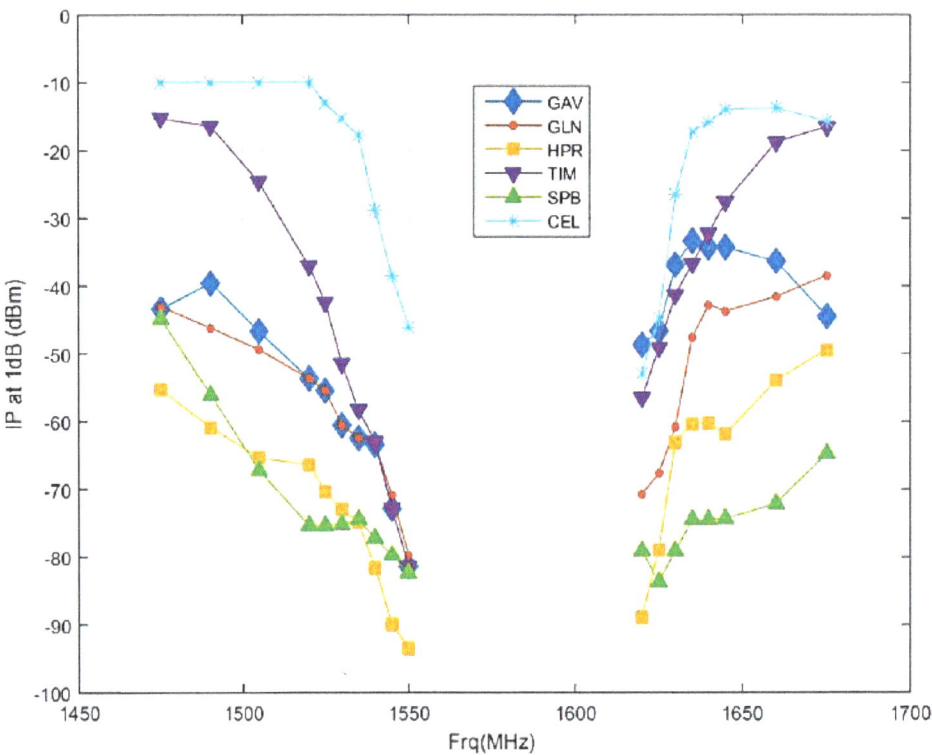

FIGURE 2-1 Bounding masks to induce no more than 1 dB of reported C/N_0 loss as functions of center frequency for various classes of GPS receiver (a repeat of Figure 1-13). The types of receivers shown are General Aviation (Non-Certified) (GAV), General Location/Navigation (GLN), High Precision (HPR), Timing (TIM), Space Based (SPB), and Cellular (CEL).
SOURCE: U.S. Department of Transportation, 2018, *United States Department of Transportation Global Positioning System (GPS) Adjacent Band Compatibility Assessment*, Washington, DC, Figure 3-22.

2.1.3 Considerations Regarding a Receiver Position Error IPC

At least for some use cases, position error is a reasonable metric in measuring harmful interference. As the primary metric of interest for many users, it is appropriate to consider accuracy. With that said, the applicability of this approach is too narrowly focused, and it will not inform harmful interference over the entire set of use cases—for example, this approach does not directly address the three other metrics of interest: availability, continuity, or integrity.

It is not possible to assess harmful interference across a range of use cases using only position error as the observable. In order to do such an assessment, testing would need to include performance assessments across GPS operational states, including cold-start and warm-start acquisition and phase-lock. Such testing would be laborious. Furthermore, the evaluation of position error must carefully consider the behavior of the receiver. The dynamic model and the method and manner of the phase measurement fusion can greatly impact the position accuracy. RTK and other differential approaches must be evaluated relative to their expected performance. Harmful interference in such use cases could be only centimeters of extra error. Additionally, some applications may experience harmful interference in manners that are not at all observable using position error—for example, applications interested in timing or velocity or scientific applications that are primarily interested in raw phase measurements. Long, perhaps prohibitive, testing would be necessary to develop statistics on mean time between cycle slips, or time-to-first-fix. These drawbacks are mitigated with SNR-based approaches.

As with the SNR-based approach, the application of uniform hard thresholds for accuracy do not make sense. Harmful impact to accuracy depends on the use case, and it should be mapped as such. The Roberson report applied an ambiguous approach to determining harmful impact based on position error.

The implementation of the position error metric also requires careful consideration. The Roberson report used a 3-minute time average to compute position error. This is not at all unreasonable when applied to some use cases. However, it would not reveal harmful interference in all use cases. To provide an example, interference might not affect average position error very much, but it might affect continuity of operations, which can be highly problematic for aviation devices or high-precision devices. Losing signals even for a very short time during a surveying operation or at an Airport Ground-Based Augmentation System receiver could cause immense operational problems for the user. If some of the lowest elevation satellites become unusable, then average position error might not increase much under standard testable conditions, but that loss might make a receiver's defenses against a spoofing attack much less robust. Therefore, simple measures of position accuracy degradation do not adequately capture all of the possible harmful interference effects that are of legitimate concern. Even if position accuracy degradation were the accepted metric of whether harmful interference had occurred, the Roberson study was inadequate for purposes of making that assessment. The testing that was completed was limited to only a subset of GPS use cases. For example, it did not consider high-bandwidth error effects. In the high-bandwidth case, error standard deviation is also important. The question of error standard deviation versus error mean is further addressed near the end of Section 2.2.4.

2.1.4 Comparison Between SNR and Accuracy Approaches

All things considered, the two approaches have more in common than not. Table 2-1 compares the approaches across a number of considerations. Both approaches are inherently device-by-device measures of harmful interference that provide little insight into the physical mechanism within the receiver that caused the degradation. Besides the effort required to conduct the device-by-device testing, such testing opens questions such as the set of devices to test, and where to draw the line on the acceptable level of harmful interference. Is the interference considered harmful if any of the tested devices

TABLE 2-1 Comparing Approaches

	Loss in Effective C/N_0	Degradation of Position Accuracy
Receiver-independent metric	No	No
Mix in-band and adjacent-band interference	Yes	Yes
Subjective thresholds	Yes	Yes
Measurable and repeatable for a given receiver design using simulators	Yes	Yes
Measurable and repeatable for a given receiver design using live sky	No	May be challenging to know the truth in regions where GPS is denied
Informs impact over a range of operating conditions	Yes; informs foundational performance: cold-start, warm-start, hot-start, code-lock, carrier lock, etc.	Not as proposed
Evaluate timing receivers	Yes	No
Informs accuracy, availability, continuity, integrity	Yes	No
Trustworthy measurement	Most estimate C/N_0 in a standard method; however, receivers' algorithms unknown	Yes

NOTE: Color coding scheme: red highlights undesirable metric properties; green highlights desirable properties; and yellow highlights lie between being desirable and undesirable.

experiences harm, or is the criterion that the "median" receiver must experience harm, or is the criterion that all receivers must experience harm? Even if all parties agree on which of the two approaches to use, considerable disagreements are likely to remain with the implementation of the approach. Decisions regarding the proper path loss model, or the acceptable impact region size, or which thresholds to use, or which receivers should be guaranteed to maintain acceptable performance would still need to be settled. How much SNR margin do receivers need? What is an acceptable frequency for cycle slips? The list goes on, and only through dialog among interested parties can these multitude of concerns be addressed.

The primary distinction between the two approaches is their predictive power. The SNR-based approach informs link margin. The link margin can forecast performance across most, if not all, of the receivers' operating modes. This simplifies testing to a large degree, but it does not make the testing simple. Moreover, with the SNR approach not only are pass/fail criteria known, the remaining margin is also known. The limitations of the SNR-based approach contemplated in Task 1 are the implicit usage of a single, firm

IPC threshold and the implicit dependence on receiver RF front-end design. Harmful interference occurs at a wide range of SNR degradation levels depending on receiver and use case, and a wide range of SNR degradation levels occurs in different receivers for the same Ligado transmission power. The receiver position error method of evaluation, on the other hand, has little predictive power under any circumstances beyond the specific use cases and receivers covered in the tests. Had the question been posed in a way that allowed for the nuances of the SNR approach and with some recognition of the applicability of receiver standards, the committee likely would have chosen SNR degradation as the clear favorite for deciding whether or not Harmful Interference had occurred.

2.2 TASK 2: HARMFUL INTERFERENCE TO GPS AND MOBILE SATELLITE SERVICES

This section considers the second task of the committee:

> The potential for harmful interference from the proposed Ligado network to mobile satellite services including GPS and other commercial or DoD services, including the potential to affect DoD operations, and activities.

2.2.1 Overview

The evaluation of harmful interference on services using the MSS/GPS band has been thoroughly investigated. The conclusions presented here are based on the available test data, analysis, and reports.

For GPS, several sets of interference tests have been performed that span many representative GPS devices drawn from many different receiver classes and suppliers. The tests evaluated various scenarios and advocated for different metrics to determine the onset of harmful interference. Despite these differences, the results consistently indicate that a majority of the devices tested do not experience harmful interference.

Based on the results of tests conducted to inform the Ligado proceeding, most commercially produced general navigation, timing, cellular, or certified aviation GPS receivers will not experience significant harmful interference from Ligado emissions as authorized by the FCC. High-precision (HP) receivers are the most vulnerable receiver class, with the largest proportion of units tested that experience significant harmful interference from Ligado operations as authorized by the FCC.

The committee found that it is within the state-of-the-practice of current technology to build a receiver that is robust to Ligado signals for any GPS application and that all GPS receiver manufacturers could field new designs that could co-exist with the permitted Ligado signals and achieve good performance even if their existing designs cannot.

For mobile satellite services (MSS), Globalstar is unlikely to experience harmful interference (only the uplink is in the L-Band, and it is a code-division multiple access [CDMA] signal). Iridium terminals will experience harmful interference on their downlink caused by Ligado user terminals operating in the UL1 band within a significant range of a Ligado emitter—up to 732 m. (See Section 2.2.5.)

2.2.2 Impact

The difficulty of determining a metric to reliably quantify the impact of interference is owing in part to GPSs being used in different ways by the communities that need higher precision. The most common use case is a roving user with modest accuracy needs. Such a user will observe interference intermittently, if at all. The most severe harmful impact is the persistent loss of code-lock tracking on multiple satellites, although this will be very rare. Yet, many users will be impacted in other ways. Stationary users may be harmed by loss of signal acquisition. High-precision users depend on stable phase-lock. Each of these use cases is impacted at significantly different interference levels.

As licensed, some aviation receivers may experience harmful interference. The conditions of FCC Order 20-48 stipulate that Ligado cannot operate a downlink (D/L) antenna at "any location less than 250 feet laterally or less than 30 feet below an obstacle clearance surface established by the Federal Aviation Administration (FAA)." Certified aviation receivers will not experience harmful interference outside of this exclusion zone, and fixed-wing manned aircraft operating under standard flight rules will never enter these cylinders of exclusion.[2] Other aviation users, however, cannot be guaranteed to remain outside these cylinders. Public safety helicopter operations, which are conducted by medevac providers, law enforcement agencies, firefighting departments, the U.S. Coast Guard, the Army National Guard, and other entities, often work in close proximity to infrastructure during the course of their operations. Crews do their very best to mitigate risk and ensure safety through a variety of methods, including visual separation and reliance on important installed safety systems, such as GPS and radio altimeters. After Ligado deployments, helicopter operations will continue to operate close to towers and other infrastructure as required by the specific mission or operation. While most such activities will operate under Visual Flight Rules within 250 feet of a Ligado tower, no guarantees can be made for GPS-only navigation within the 250 × 30 feet cylinders.

---

[2] E. Drocella, C.-W. Wang, and N. LaSorte, 2020, "Assessment of Compatibility Between Global Positioning System Receivers and Adjacent Band Base Station and User Equipment Transmitters," NTIA-TM-20-536, Section 5.2.3.9, U.S. Department of Commerce.

The same can be said for GPS-only navigation by unmanned air vehicle systems that will provide critical services in the future. Such systems may experience harmful interference to their GPS navigation subsystems when inside the cylinders and must rely on other sensor input, including cameras. Ligado interference may occur exactly when GPS is informing a critical safety service, such as GPS-based terrain avoidance after an unexpected transition to low visibility conditions. Supplementary navigation instruments, including radar and fixed beaconing, may mitigate the occasions when visual flight rules cannot be relied on and proximity to a Ligado tower results in loss of GPS, but there is presently no guarantee of this.

The accuracy of high-precision applications is complicated and cannot be reduced to a simple assessment of the impact of carrier-to-noise density ratio, or the impact of a 1 dB reduction in that ratio.

2.2.3 Test Results

To assess interference with adjacent channel Ligado signals on GPS performance, key tests were performed by four different groups:

- Roberson and Associates (RAA), a consulting firm sponsored by Ligado;
- National Advanced Spectrum and Communications Test Network (NASCTN), a confederation of the federal spectrum and testing experts, commissioned by Ligado;
- U.S. Department of Transportation (DOT), which created the Adjacent Band Compatibility (ABC) report; and
- A DoD classified assessment that was viewed by a cleared subset of the committee.

The DOT ABC assessment included two primary components, one led by the DOT Office of the Assistant Secretary for Research and Technology and focused on civilian GPS devices and their applications, and the other led by the FAA and focused on certified GPS avionics.

Appreciating the Difficulty of Testing

These testing studies came up with varying and conflicting assessments of interference potential, which is not unreasonable given the complexity of the assumptions, differences in metrics, and experimental processes. Each testing study has its relative strengths and weaknesses, and they are outlined below, along with key results. Testing is an essential part of the regulatory assessment; it does help in reducing, but not eliminating, uncertainty regarding underlying assumptions or circumventing the need for all the details.

Testing has its limitations. Not all makes and models of GPS receivers produced can be tested and a representative set is needed. Those GPS receivers selected must be tested under fair, realistic conditions. Metrics depend on what needs to be tested and how that metric relates to harmful interference. The FCC regards harmful interference as interference that impacts the functionality of a device; again, functionality is a subjective measure. In the case of GPS, functionality is the performance of the GPS device, which is somewhat subjective and use-case dependent. Thus, identifying a single metric to characterize Harmful Interference is the most challenging part of creating a non-controversial testing process.

Although all tests looked at the C/I (or C/N_0) metric, this metric is self-reported from the device, and the measurement is non-standardized in concept and in hardware implementation, making it a problematic metric. (See Section 2.1.) Variations of up to 62 dB in device-reported C/I were measured between devices under the same conditions in the ABC study.

Furthermore, applying the test results to a relevant operational scenario requires the selection of a propagation model to determine the range at which the observed effects will occur. These models are highly dependent on the geometry of the scenario and can vary by many tens of decibels. A conservative approach uses free space propagation loss which is useful for cases where there is a clear line of sight (LOS) between the transmitter and the receiver. A non-conservative approach is to use a non-LOS model, which would occur if the receiver were masked from the transmitter by buildings or terrain. Both of these approaches as well as approaches that combine the two (LOS for distances < ~100 m and non-LOS for longer distances with a smooth transition between) were used in the various reports.

The committee realizes that all testing is subject to limitations, but that testing can be used to evaluate the possibility of harmful interference if appropriate post-test analyses are applied. One obvious analysis is to determine equivalent stand-off distances under a given propagation model. It would be impractical to test all aspects of all significant use cases and modes of harm to receiver operations, but it is possible, in theory, to deduce when the various modes of harm will occur in the various use cases based on receiver test data of a more limited nature. In order to enable such analyses, it is necessary to provide the raw (post calibration) data from all of the tests;[3] only the NASCTN "LTE Impacts on GPS" tests did this. In addition, no guidance was given about which receivers should be protected from harm. Should it be all receivers tested, only those with a certain market penetration, only those serving highly important applications, or only those with the best RF front ends? Therefore, despite all the tests, the available data

---

[3] Dependent variables were self-reported by the devices under test, which are dependent on the receiver point design.

and analyses fell short of what the committee would have liked to see in order to make the best possible decision about whether the permitted Ligado tower transmission levels would cause unacceptable harm to GPS receivers. Nevertheless, the committee was able to draw some conclusions about harmful interference based on the test data, on rudimentary analyses of its own, and on information provided by various entities that gave presentations to the committee.

It is important to note that no mobile satellite systems interference tests were presented to show the impact on systems such as Iridium. Iridium presented an analysis. Several members of the committee produced their own independent calculations that verified the correctness of the core finding of the Iridium analysis: that interference to its downlink signals will occur. DoD provided a classified test report from 2016 to quantify the impact of Ligado interference on their systems.

Overview of Test Approaches and Results

The test objectives, setup, limitations, and relevance to FCC Order 20-48 are summarized in Table 2-2.

Both the NASCTN and RAA reports did show that some high-precision receivers could be adversely impacted by Ligado emissions. The RAA study showed that high-precision receivers it tested could be fixed by replacing the antenna with one that is more frequency selective (when replacing an external antenna is possible), but the testing did not include HP functionality, such as RTK. For some platforms, such as weapon systems or aircraft, replacing an antenna may not be feasible. Military systems often leverage the L2 GPS signal, which will make it much more robust to interference if the additional L2 reception has been incorporated primarily as a means of increasing the receiver's robustness to jamming. If, on the other hand, a particular application requires both the L1 and L2 signals in order to remove the ionosphere error term, then the use of L2 will not mitigate any harmful interference to the L1 GPS signal that might be caused by Ligado.

2.2.4 GPS

The radionavigation satellite services (RNSS) band ranges from 1559–1610 MHz and contains the GPS L1 band. There are several sources of interference that can fall into this band, as illustrated in Table 2-3.

GPS In-Band Interference

Noise levels in the GPS band are approximately −174 dBm/Hz, so at a distance of 10 meters from a Ligado transmitter limited to −130 dBm/Hz in the GPS band (56.5 dB free space path loss at 1584.5 MHz), the interference from the Ligado user or base station will be at least 12 dB below the noise floor.

TABLE 2-2 Tests Relevant to FCC Order 20-48

	ABC	NASCTN	Roberson and Associates
Sponsor	U.S. Department of Transportation (2018).	Ligado (2017).	Ligado (May and June 2016).
Metrics	C/N_0 degradation, acquisition time C/I.	Carrier-to-noise density ratio (C/N_0), 3D position error, timing error, number of satellites in view, time-to-first-fix (TTFF), and time-to-first-reacquisition (TTFR).	C/N_0, position error.
Test Objective	Determine the conditions where C/N_0 loss would exceed 1 dB from adjacent channel signals. The report also included (1) an analysis of impact on certified aviation receivers and (2) results on acquisition time at 1 dB degradation, including for low-elevation satellites.	Testing for GPS performance in addition to power levels.	Testing for GPS performance in addition to power levels. Sought to analyze C/N_0 applicability.
Devices Tested	80 GPS receivers were tested simultaneously. Cellular, general navigation, high-precision, timing, general aviation (non-certified), space-based receivers.	14 in different categories such as general navigation, high-precision, GPS disciplined oscillators.	27 devices of various receiver classes. General navigation (2D position error), high-precision devices, 3D positioning error. The 3rd Generation Partnership Project performance testing was performed for accuracy, sensitivity, dynamic range, motion testing.
Experimental Setup	All of the devices observing GPS and LTE signals from fixed overhead antennas in a semi-anechoic chamber.	Tested receivers one at a time with the same conditions. Assumption of free space propagation losses.	GPS simulator used in a calibrated anechoic chamber. Ligado uplink and downlink tested with power levels corresponding to a fully loaded system (rare in real systems). Maximum out-of-band interference created in the GPS band while in-band power varied.

TABLE 2-2 Continued

	ABC	NASCTN	Roberson and Associates
Test Limitations	Relied on C/I metric. All the devices observing GPS and LTE signals from fixed overhead antennas, each having a different relative orientation, making comparisons difficult. Potential issues with electromagnetic interference between devices are also unknown and could explain the time-varying noise floor observed. Did not make the data available for study to assess normal C/N_0 variations.	Not as many devices tested.	Could have considered additional key performance indicators (KPIs). No consideration made for acquisition or phase tracking performance.
Relevance to FCC Order 20-48	This study appears to be discounted by the FCC because the 1 dB C/N_0 metric was not linked to harmful interference for GPS. Furthermore, the 1dB C/N_0 was applied to the adjacent band, not the in band, which breaks with past FCC precedent in using this metric. Last, 2 dB variations in the noise floor, even without Ligado, have been observed for certain orbits and geometries.	Showed that in a few cases high-precision receivers can be impacted. Unclear if these receivers have a significant number, especially those still in operation given their age and inability to exploit the newer GNSS constellations. Variations in testing from 2 to 3 dB C/N_0 were observed without Ligado interference.	Testing showed no correlation between 1 dB C/N_0 degradation and position errors. Demonstrated mitigation strategies using modified external antennas. RAA claimed no impact on any receiver from co-channel interference. No impact to the general navigation receivers except in one case. Of the 11 high-precision units, 4 had no impact in 3D error. The others showed impact owing to the 1526–1536 downlink when at maximum power, with performance degradations occurring from −25 to −55 dBm, depending on the GPS unit. Three were remedied by more selectivity at the antenna. The other 4 could not change out their antennas. Of the 11 high-precision units, three of the receivers were impacted by the 1627.5–1637.5 MHz uplink channel, with two fixed by improving the selectivity at the antenna. One receiver did not allow for an antenna change and was left impacted by interference. Tests showed random variations in C/N_0 in excess of 1 dB with no interference present, if no averaging was used.

TABLE 2-3 Sources of Interference in GPS Band

Source	Frequencies (MHz)	Power (dBm/Hz)
Ligado user	1559–1608 and 1608–1610	−135 Linear −135 to −130
Ligado base station	1559–1610	−130
AWS-3	>1695	−73
Iridium MSS	1617.775–1626.5	
Globalstar MSS	1610–1618.725	
Ligado MSS	1525–1559 and 1626.5–1660.5	

For the case of multiple base stations, the separation requirement of 433 m ensures that a given GPS device is likely only to be impacted by a single Ligado base station. For the multiple user case, four users would need to be within a 10 m radius of the GPS receiver and be transmitting simultaneously to reduce the GPS C/N_0 by the 1 dB advocated for by the GPS vendors.

As a result, the in-band interference allowed by FCC Order 20-48 users or base stations are unlikely to have any adverse effect on GPS operations.

GPS Out-of-Band Interference

Out-of-band emissions (OOBE) near GPS (less than 1559 MHz or greater than 1610 MHz) can potentially impact GPS receiver performance. The degree of interference will depend on several receiver design factors, including the receiver front-end bandwidth and dynamic range, the sampling rate, and signal processing used. The power, frequency, directionality, and duty cycle of the interferer are also critical parameters.

In existing GPS receiver designs, aliasing of out-of-band signals into the GPS band is typically not a problem. However, the large power discrepancies between the Ligado base station/user and the GPS signal make it possible for aliased power to be folded into the GPS band and may adversely affect receivers not designed for that environment. The center of the RNSS band is at 1584.5 MHz, the Ligado base station downlink is at 1531 MHz, the lower Ligado user uplink is at centered 1632.5, and the upper Ligado user uplink is at 1651.5 MHz. The frequency difference between the center frequencies of the Ligado signals and the RNSS band are 53.5 MHz, 48 MHz, and 67 MHz for the Ligado downlink, lower uplink, and upper uplink respectively. A GPS receiver with a sampling rate below twice the difference frequency must employ adequate filtering before the A/D converter to avoid aliased interference. This design constraint is normal for any digital system. As an example, a GPS receiver at a distance of 10 m (56.5 dB path loss) from a base station (39.8 dBm), would require an attenuation of 111 dB to be at the same level as the GPS signal (−128 dBm), which would need to be provided by the analog filtering in the front end.

Note that the exact frequency plan (intermediate frequency [IF], sampling rate, number of down-conversion stages) for a particular GPS receiver is a critical design parameter, and receiver design depends heavily on the spectral environment and the use of the receiver. GPS receivers operating in a quiet environment will require less attenuation, while receivers that use corrections delivered via an MSS signal will need to receive those frequencies as well. The current Ligado downlink band (1526–1536) was used to deliver these corrections, but they have moved several times up to 2016, when Ligado agreed to permanently locate them in the 1555–1559 MHz range.[4] As such, the Ligado downlink band is within the passband of some legacy receivers.

In summary, dealing with OOBE is a standard design problem for digital receivers and can be accommodated through a number of well-understood design techniques as long as the environment is well characterized and understood. The committee found that it is within the state-of-the-practice of current technology to build a receiver that is robust to Ligado signals for any GPS application, and several manufacturers have done so. The GPS/MSS band environment has been in constant flux since the introduction of the Ancillary Terrestrial Component (ATC) rule, making it difficult if not impossible for receiver vendors to anticipate the environment a receiver will encounter over its lifetime. Furthermore, every rule change risks making obsolete equipment with design assumptions that were violated by the change.

Test Results

C/N_0

Several tests were designed to evaluate the degradation in C/N_0 that occurs as a result of an interfering signal. The plots in the figures below derive from National Telecommunications and Information Administration (NTIA) TM-20-536, which analyzed data from several previous test events to inform the community regarding the proposal that was being considered in the 2020 time frame (and largely manifested itself in FCC Order 20-48). The data constitute an aggregate from tests performed by the DOT Adjacent Band Compatibility, DoD, Roberson and Associates, National Advanced Spectrum and Communications Test Network, 2011 FCC Technical Working Group, and 2012 National Positioning, Navigation, and Timing Systems Engineering Forum. The data look across a large population of receivers in several different categories, high precision (HP), general

---

[4] S. Riley, Trimble, Inc., 2021, "Trimble Presentation to NAS," Presentation to the Committee to Review FCC Order 20-48, January 12, 2022, Washington, DC: National Academies of Sciences, Engineering, and Medicine.

location/navigation (GLN), timing (TIM), cellular (CEL), non-certified general aviation (GAV), and space based (SPB) that were evaluated with varying scenarios in the different studies.

Plots showing the percentages of receivers that experienced 1 dB, 3 dB, and 5 dB degradation in C/N_0 at each interferer power level are shown for the HP, GLN, TIM, and GAV classes for a single base station (DL) or a single user handset (UL1/UL2). In addition, vertical range lines indicate the levels that would be experienced by a receiver at 1 m, 10 m, and 100 m from the interferer using a standard free space loss model, $20 \log_{10}(4\pi d/\lambda)$ at the center of the GPS band. These reference lines do not include antenna gain/polarization losses (typically ~5 dB), which are situation dependent and can be accounted for by translating the lines to the left. Note that the number of receivers in each case varies based on the tests that were performed in each of the campaigns from which those data were aggregated. For uplink cases, no power control was assumed, which is indicative of the case where the user terminal is on the periphery of the coverage area. This is conservative, but the area in a coverage region is concentrated in the regions farthest from the base station. The impact of power control for a particular user can be considered by scaling the interferer power axis for a given range to the base station.

Ligado Base Station Interferer

Figure 2-2 is for a single Ligado DL interferer in the 1525–1535 MHz or 1526–1536 MHz band, depending on the study the interferers were drawn from. The range lines use a 9.8 dBW base station power per FCC Order 20-48.

Ligado UL1 Interferer

Figure 2-3 is for a single Ligado UL interferer in the 1627.5–1637.5 MHz band. The range lines use a –7 dBW user power per FCC Order 20-48.

Ligado UL2 Interferer

Figure 2-4 is for a single Ligado UL interferer in the 1646.5–1656.5 MHz band. The range lines use a –7 dBW user power per FCC Order 20-48.

In looking through the cumulative distribution plots, many GPS receivers that were tested do not show a significant degradation in C/N_0 even at very close ranges. There are, however, many GPS receivers that show 1–5 dB degradation even at ranges beyond 100 m. FCC Order 20-48 allows a deployment with 433 m between base stations or a maximal range of 250 m to a base station. On the uplink side, mobile Ligado users can be

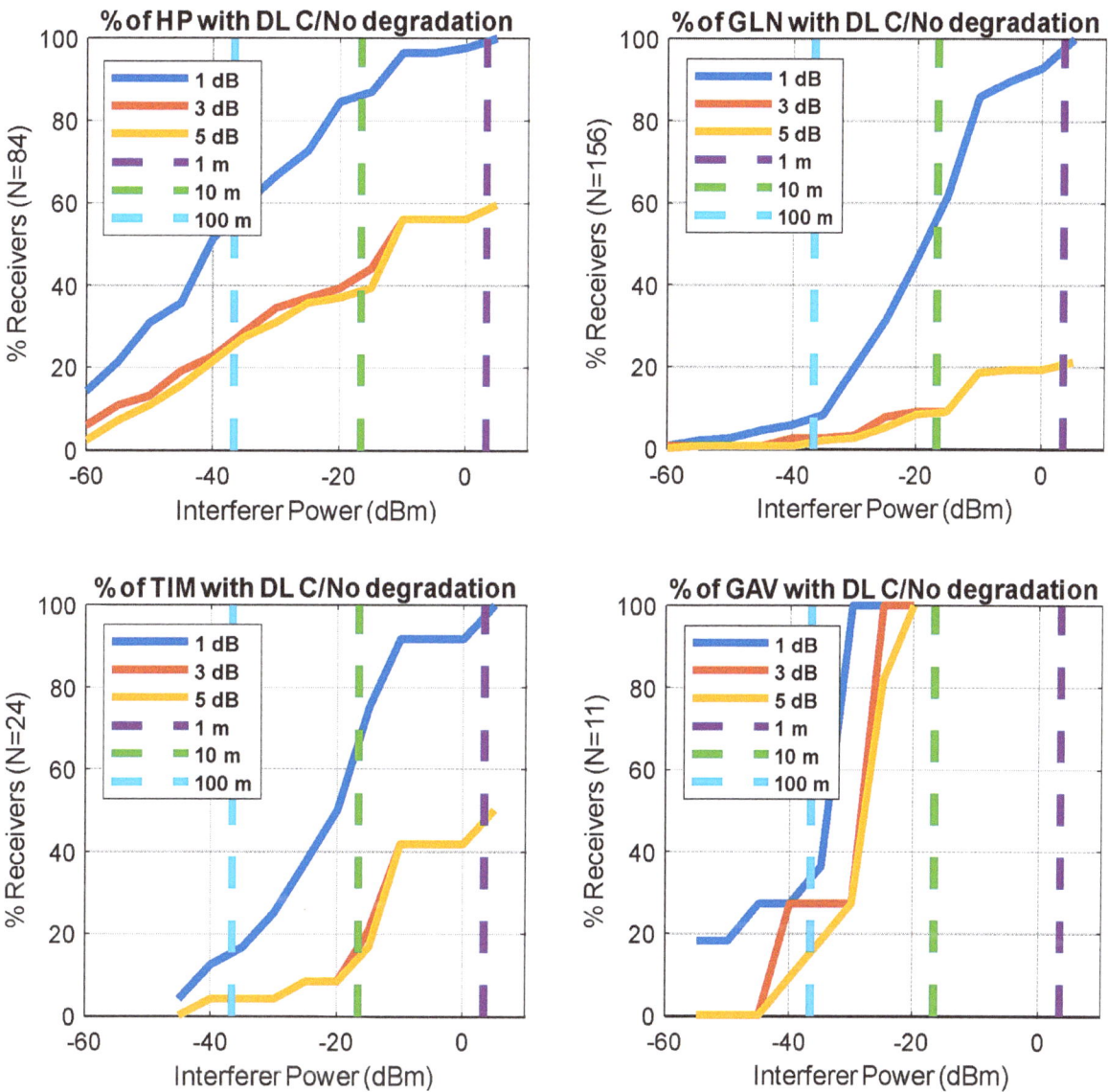

FIGURE 2-2 Percentages of four classes of receivers that experience three levels of C/N_0 degradation as functions of a single Ligado downlink interferer's power in either the 1525–1535 MHz band or the 1526–1536 MHz band.

numerous, and can be anywhere in the coverage area, each contributing to interference; however, interference will be dominated by the closest user(s), owing to the path loss.

Mean and Standard Deviation of Position Error

In addition to C/N_0, several studies have evaluated the use of the position error as a metric to determine harmful interference. The RAA study evaluated 2D position error, and the NASCTN study evaluated 3D position error. Figure 2-5 shows the mean and mean +3 standard deviations of the data overlaid on a scatter plot of the 3D position error versus the

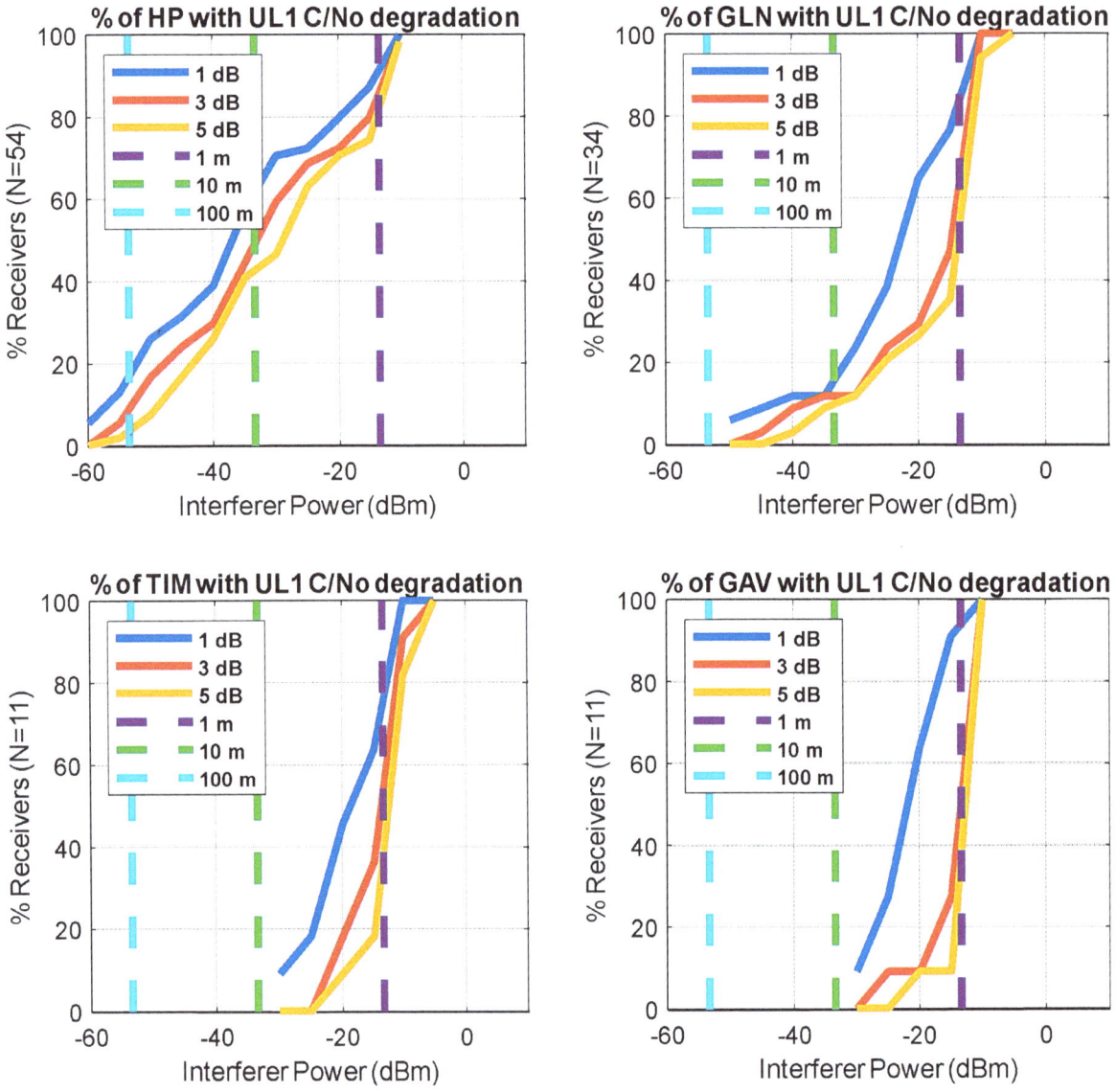

FIGURE 2-3 Percentages of four classes of receivers that experience three levels of C/N_0 degradation as functions of a single Ligado uplink interferer's power in the 1627.5–1637.5 MHz band.

interferer power level at the GPS receiver using data from the NASCTN study. In addition, vertical lines indicate power levels that correspond to ranges between the receiver and the interferer of 100 m, 10 m, and 1 m as calculated using free space path loss. As an example, a GLN device (DUT2) is shown on the left. In this particular case, the spread of the position error, the mean, and the standard deviation of the position error are relatively stable with respect to the interference power level. The figure on the right shows a similar plot for a high-pressure processing (HPP) device (DUT7). In this case, the mean remains relatively flat, but the spread increases by more than a factor of 4 at a power level that is consistent with a range of greater than 100 m, indicating degradation owing to interference.

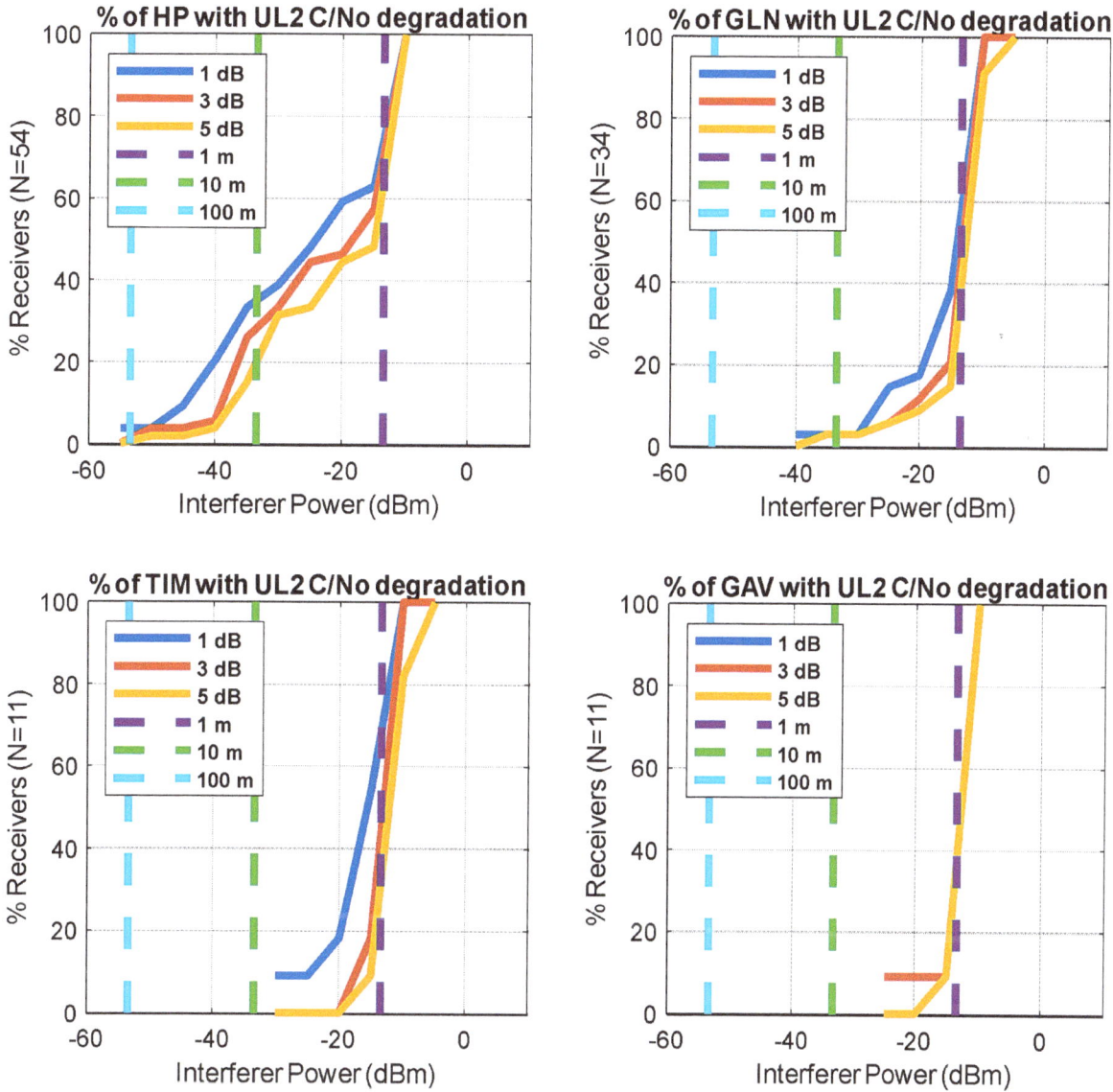

FIGURE 2-4 Percentages of four classes of receivers that experience three levels of C/N_0 degradation as functions of a single Ligado uplink interferer's power in the 1646.5–1656.5 MHz band.

The NTIA TM-20-536 report looks at similar parameters from the NASCTN and research and analysis studies and additionally correlates the position error with receiver-reported C/N_0. The conclusion drawn from the aggregate of all of the data is that the mean of the position error does not indicate the increased spread in the positioning error whereas the standard deviation provides a more suitable indicator of the impact of the interference on the quality of the data provided by the device. Therefore, the mean of the position error should be augmented with additional error statistics (standard deviation, root mean square [RMS], sample error distribution, etc.) in the evaluation of harmful interference to capture the increased spread in position error with increased interference.

Similar plots for UL1 and UL2 for these same devices appear in Figure 2-6.

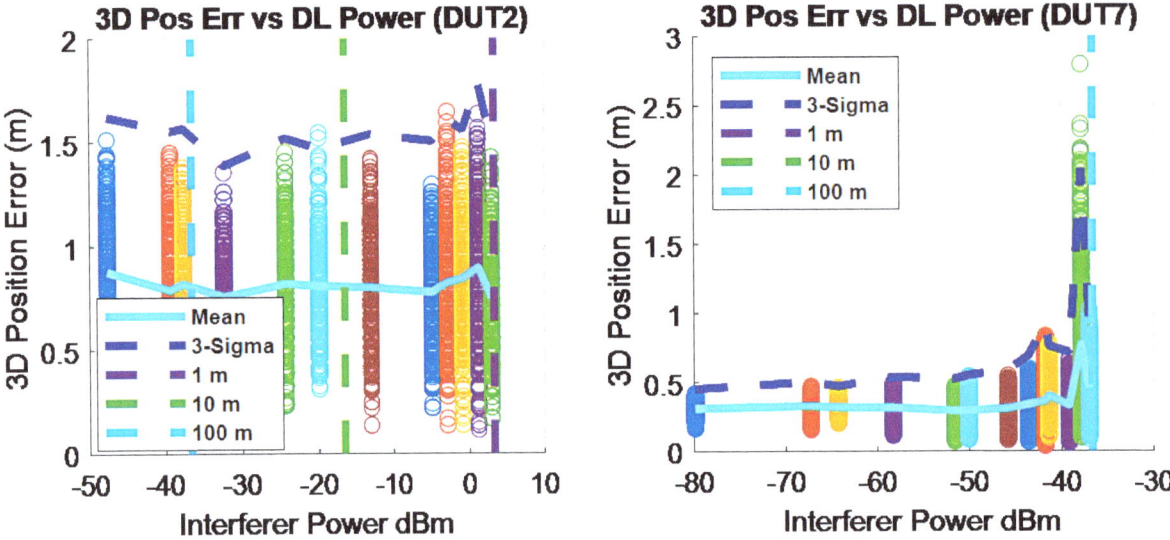

FIGURE 2-5 3D position error versus interferer downlink power level overlaid with mean and mean +3 standard deviations for two devices under test in the NASCTN study.
NOTE: DUT2 (left) is a GLN device, and DUT7 (right) is an HPP device.

2.2.5 Mobile Satellite Services

The MSS band supports uplinks and downlinks between mobile users and satellite relays. (See Figure 1-2.) The band, discussed above in Section 1.1.2, is currently divided into three sections: GEO downlinks (1525–1559 MHz), GEO uplinks (1626.5–1660.5 MHz), and Big LEO uplinks/downlinks (1610–1626.5 MHz). GEO Earth terminals typically use high-gain antennas to compensate for the long range and thus have natural attenuation of signals that are not line-of-sight to the satellite. The focus here is therefore on the impact of ATC OOBE on the Globalstar and Iridium systems.

Earth to Satellite

Interference on the satellite uplinks is mitigated by limiting Ligado OOBE into the uplink bands and through the radiation pattern, which directs energy toward the terrestrial users instead of into space.

The Globalstar system's uplink transmissions are at 1610–1618.725 MHz, and use a CDMA waveform that inherently provides some immunity to interference.[5] The downlink is in the C-Band. The frequency plan of the Globalstar system coupled with the levels of the ATC OOBE (−130 dBm/Hz at 1610 MHz to −101 dBm/Hz at 1618.725) into the Globalstar uplink band do not result in interference at the satellite from the ATC emissions.

---

[5] Globalstar, Inc., 2017, "Globalstar Overview," https://www.globalstar.com/Globalstar/media/Globalstar/Downloads/Spectrum/GlobalstarOverviewPresentation.pdf.

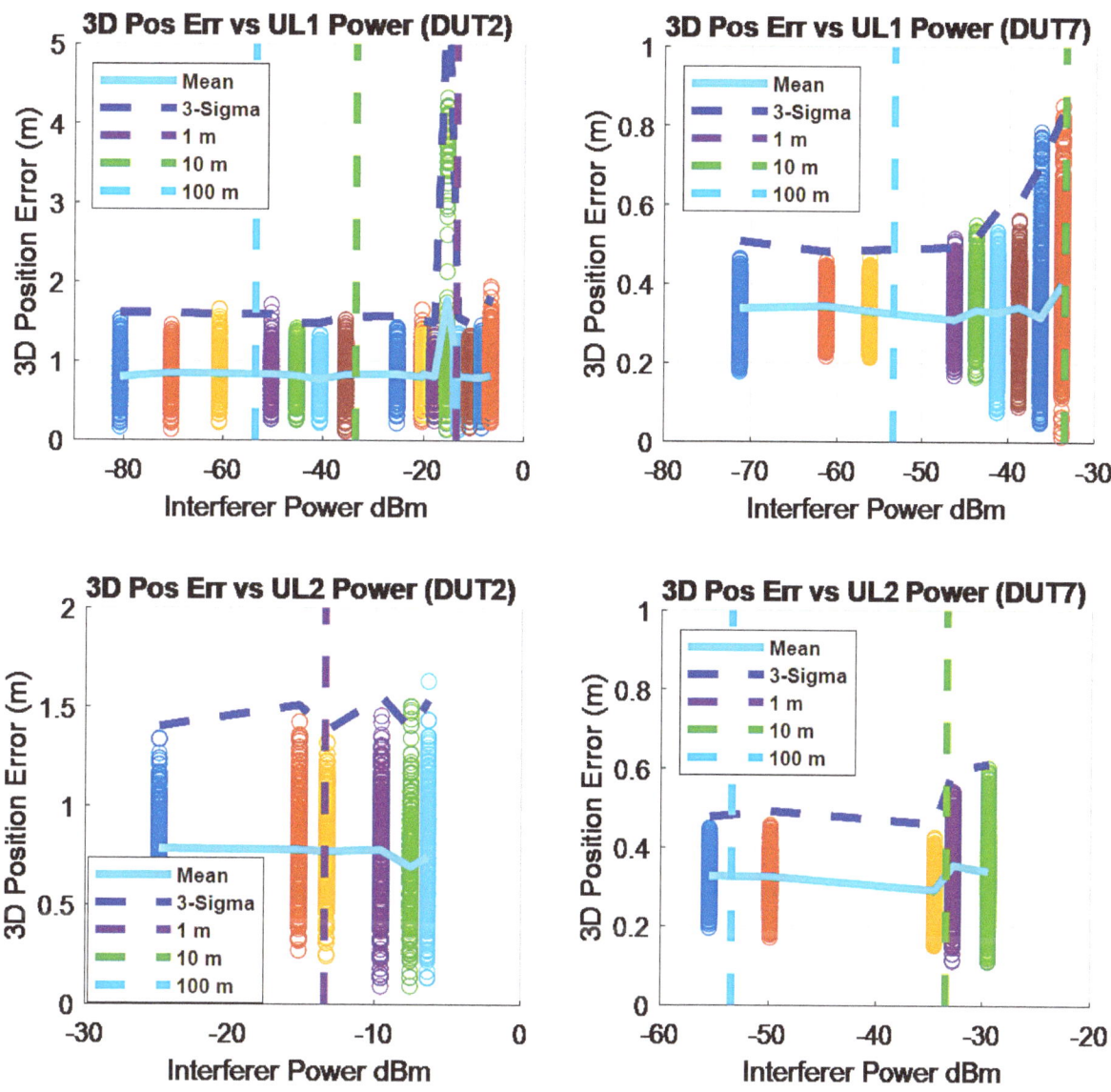

FIGURE 2-6 3D position error versus interferer uplink power level (in uplink bands 1 and 2) overlaid with mean and mean +3 standard deviations for two devices under test in the NASCTN study.

The Iridium system uses a frequency-division multiple access/time-division multiple access multiplexing scheme, so the uplinks and the downlinks share the frequencies from 1617.775–1626.5 Mhz. The ATC OOBE into the Iridium band are −104.7 dBm/Hz at 1617.775 MHz to −76.26 dBm/Hz at 1626.5 MHz. Similar arguments can be made for the Iridium uplink as for the Globalstar uplink and despite the higher interference power levels at the satellite, no claim of interference has been made.

Satellite to Earth

The Iridium downlink shares the same frequencies as the uplink and is subject to the same interference. Iridium has stated their noise floor is −170 dBm/Hz, which is consistent with a noise-limited system.[6] Interference levels that approach this level will have an impact on Iridium performance. Unlike the uplink case, the radiation pattern will direct the energy toward the Iridium terminal.

The OOBE spectral mask for ATC users is a linear ramp from −73 dBm/Hz at 1627.5 MHz to −130 dBm/Hz at 1610 MHz followed by a linear ramp to −135 dBm/Hz at 1608 MHz, as shown in Figure 2-7. In addition, the OOBE from the upper uplink band is −64 dBm/Hz at 1645.5 MHz with no clear lower bound until the −73 dBm/Hz restriction at 1627.5 MHz. When multiple users are present, the spectral masks from each user will be additive after being adjusted for the range of each user to the Iridium receiver.

For the Iridium band, channels on the high side will see an interference level of −76 dBm/Hz from a single user, which will require 94 dB of attenuation to be reduced to the noise floor. This will occur at a distance of 732 m (free space path loss) or 51 m (non-line-of-sight).[7] On the lower end of the band, the interference level is −102 dBm/Hz, which will require 68 dB of attenuation to be reduced to the noise floor. This will occur at a distance of 40 m (free space path loss) or 16 m (non-line-of-sight).

The committee's conclusion is that the Iridium downlink will experience harmful interference at relevant ranges. The above analysis assumes a single user; the situation will be both more likely and more severe as the spatial density of the users increases.

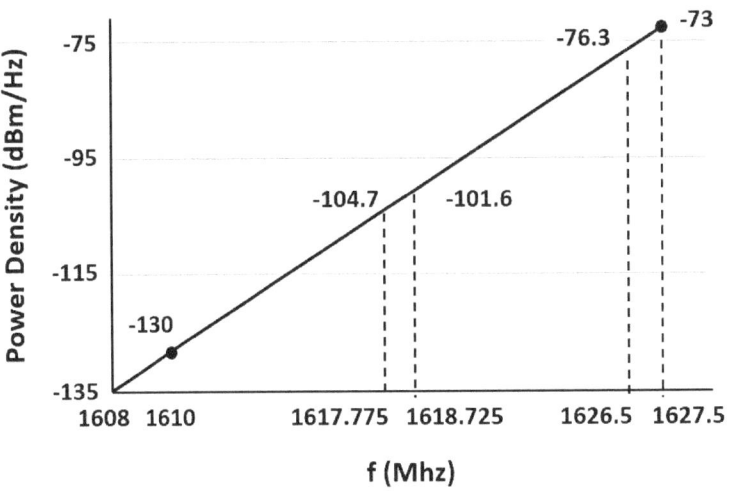

FIGURE 2-7 Out-of-band emissions spectral mask for ATC users.

---

[6] Iridium Communications, Inc., 2020, "Petition for Reconsideration," *FCC Proceedings* IB 11-109 and IB 12-340, Table 3, May 22.

[7] Non-LoS range determined using the COST 231 model for an urban environment with a frequency of 1626.5 MHz, Tx antenna height of 4 m, Rx antenna height of 1.5 m, average roof height of 15 m, distance between buildings of 35 m, street width of 17.5 m, and a 45-degree propagation angle.

2.2.6 U.S. Department of Defense

DoD uses the MSS band for many different applications, including communications, navigation, tracking of military assets such as equipment and vehicles (including drones), synchronization of timing, and so on. In addition, the U.S. government is by far the single largest Iridium customer.

GPS can be relatively easy to jam; numerous studies have indicated this. One can even find commercially available but illegal GPS jammers on the market that are used to defeat tracking systems. Military system GPS receivers are more difficult to jam. The military uses an augmented L2 signal that provides resilience to jamming around L1, if it is not needed for ionospheric corrections. (See Section 2.2.3.) Other techniques such as null-steering antennas can provide additional robustness.

DoD is also a major customer of commercial mobile satellite systems, such as Iridium. DoD has evaluated the impact of FCC Order 20-48 on department devices and missions. The following summary points were provided to the committee in a set of slides dated March 15, 2022. It is important to note that these conclusions were asserted by DoD without providing publicly available supporting data and were not discussed by the committee in a public session.

- DoD and interagency partners conducted testing to determine the impacts to GPS (captures FCC Order 20-48's authorized deployment). The tests demonstrated that the proposed signal introduces harmful interference to critical national security mission capabilities.
- The terrestrial network authorized by FCC Order 20-48 will create unacceptable harmful interference for DoD missions. The mitigation techniques and other regulatory provision in FCC Order 20-48 are insufficient to protect national security missions.

In a subsequent response to questions from the committee, DoD also commented that there were no independent studies or technical analysis of Iridium's claims of interference, and that although DoD did initiate its own analysis of some of the claims, that work was not completed as essential information was unavailable to DoD.[8]

Additional information on the test results and analysis as they related to DoD systems and missions is discussed in a classified annex to this report.

---

[8] DoD responses to committee questions, April 29, 2022.

2.3 TASK 3: FEASIBILITY, PRACTICALITY, AND EFFECTIVENESS OF MITIGATION MEASURES IN THE FCC ORDER

Task 3 concerns the feasibility, practicality, and effectiveness of the mitigation measures proposed in the FCC Order 20-48 with respect to DoD devices, operations, and activities.

2.3.1 Overview

The FCC Order 20-48 enumerates several potential mitigations when a receiver experiences harmful interference. These include:

1. Mandating exclusion zones for Ligado emitters (as was done for potential navigation receivers in the order).
2. Replacing components of vulnerable receivers (e.g., replacing antennas in some high-precision receivers was discussed in several reports).
3. Replacing all vulnerable GPS receivers.
4. Moving receivers farther away from Ligado emitters when practical.
5. Putting a regime in place to require Ligado to have a facility to turn off emitters in some geographic locations on notice.
6. Negotiated mitigations between Ligado and the affected government agency to determine "an acceptable received power level over the military installation."

In addition, the FCC requires Ligado to maintain a database of Ligado emitters. The database will be helpful in ruling out Ligado's influence in harmful interference in many cases and help identify potentially interfering Ligado emitters, but attributing harmful interference as experienced by identified receivers to one or more Ligado emitters will be difficult.

In the case of harmful interference with GPSs, the effectiveness and practicality of any of the foregoing potential mitigations depends on the type of receiver and the application. It is important to distinguish between two types of equipment:

1. *DoD Authorized/Compliant Devices* approved for weapons and weapons delivery systems and other national security certified devices.
2. *Commercial GPS Devices* when used in national security applications with an express waiver or navigational warfare (NAVWAR) compliance determination per DoD Instruction 4650.08 and Chairman of the Joint Chiefs of Staff Instruction (CJCSI) 6130.01 or commercial devices that are used in other DoD operations or missions such as emergency response or partner operations.

In its response to Task 2, the committee assesses that harmful interference with commercial devices is substantially unlikely with respect to CEL, GLN, and TIM devices as well as HP devices sold after circa 2012.[9] The committee also believes that it is reasonable to assume that DoD Authorized/Compliant Devices are expected to withstand willful interference under substantially higher power than is authorized under FCC Order 20-48. However, no specific information in this regard was presented or made available as part of the committee's public study.

Notwithstanding this, the analysis below evaluates the potential mitigations in the context of device type and application below.

In addition, aside from GPS applications, DoD operations may employ Iridium services that may be affected by Ligado uplink emissions near 1627.5 MHz, as is discussed below. Only potential mitigation option 6 above is applicable in this case, and the committee does not have enough information to assess its potential effectiveness or practicality.

2.3.2 Inside the United States

As noted above, the committee assumes that DoD Authorized/Compliant GPS receivers or systems that incorporate such receivers used inside the United States are built to withstand jammers emitting much higher interference power than the authorized Ligado emissions, jammers that are purposely designed to interfere with civilian and military GPS in a NAVWAR environment. While such DoD Authorized/Compliant GPS receivers are unlikely to experience degradation owing to Ligado emissions, the committee did not have access to concrete information about the actual prospect of harmful interference in these devices.

If any DoD Authorized/Compliant Devices GPS receivers or systems that incorporate such GPS receivers used inside the United States experience harmful interference, there are relatively few satisfactory mitigations because these systems must pass very long and expensive operational test certification; generally, actions that include replacing antennas or electronics to provide mitigation would involve unsatisfactorily long delays. In some cases, replacing older devices with newer versions of such devices that are protected from harmful interference and already qualified may provide a plausible solution. However, as noted above, the likelihood of such interference, especially given L2 resilience, seems very low.

The remainder of this section's comments apply to DoD's use of Commercial Devices. First, potential harmful interference to GPS receivers from Ligado emitters in

---

[9] See, for example, S. Riley, Trimble, Inc., 2021, "Trimble Presentation to NAS," Presentation to the Committee to Review FCC Order 20-48, January 12, 2022, Washington, DC: National Academies of Sciences, Engineering, and Medicine.

the 1526–1536 MHz band or GNSS band are considered, recalling that the committee judges the likelihood of such interference as low.

Replacing antenna subsystems in older commercial devices that experience harmful interference when close to a Ligado emitter can be effective. Not all vulnerable HP devices can practically be retrofitted in this way.

As pointed out earlier, if devices that may experience harmful interference can be moved farther away from a Ligado emitter, any harmful interference should be mitigated. For example, if such a receiver can be moved from operating within 3 meters of a Ligado emitter to 30 meters from a Ligado emitter, it will then experience a 20 dB decrease in interference power. Whether this is practical is application specific.

Replacing an older commercial GPS receiver with a newer one should be an effective mitigation. However, there is a cost (high in the case of HP receivers), and in some cases it may be difficult to actually substitute the devices.

When DoD employs Commercial Devices in operations that occur in fixed locations (such as a base), simply requiring that there be no Ligado emitters within 500 meters of the location should be highly effective and practical.

Last, in some cases, Commercial Devices may be replaced with alternative geolocation services, but this is likely to be expensive.

The principal mechanism identified by the FCC for mitigation requires joint exploration between Ligado and the entity experiencing a specific interference problem, and this mitigation is case specific. Additionally, attribution of GPS performance degradation to Ligado is potentially difficult to prove. When asked by the committee if DoD could describe any engagement with Ligado to resolve any specific or general interference scenarios or cases since the April 22, 2020, Order and Authorization, DoD responded, "Per NTIA direction there has been no engagement with Ligado since the April 22, 2020, Order and Authorization pending resolution of the petition for reconsideration."

The previous discussion addressed the potential harmful interference from Ligado emissions on DoD operations as it relates to GPS receivers. This analysis now turns to potential harmful interference from Ligado's uplink emissions in the band between 1627.5 and 1637.5 MHz. Uplink transmissions present no material impact to GPS service in the 1559–1610 MHz band.

A particular concern is for DoD's use of commercial Iridium services just below 1626.5 MHz. Ligado uplink emitters near 1627.5 MHz may adversely affect Iridium terminals within up to about 750 meters of such a Ligado uplink emitter. The experimental record with respect to these effects is not nearly as exhaustive as for the devices in the RNSS band, but crude calculations indicate that harmful interference is possible. These calculations are based solely on the interference power permitted by the FCC order near 1627.5 MHz. In addition to communications services, Iridium provides supplemental aviation navigation services. These should not be affected by emission from the Ligado system at altitude or when operating outside the United States.

The committee notes that there are conflicting interpretations of the regulatory requirements that apply to this potential scenario. In paragraph 117 of its order in the Ligado proceeding, the FCC, referencing a 2005 commission ruling,[10] requires services in the 1626.5 MHz band to tolerate interference substantially higher than that which would be produced by a Ligado uplink transmitter. Iridium has countered[11,12] that FCC regulations place the onus on the interferer to remedy problems caused by the permitted level. Consistent with its intent to focus on whether there is or is not harmful interference, the committee makes no judgment about the regulatory questions raised here.[13]

There are several potential mitigation options for this interference. First, it may be possible to require much sharper filtering by Ligado uplink transmitters operating near 1627.5 MHz. Second, it may be possible to bar Ligado uplink emitters within, say, 1,000 meters of potentially affected DoD operations inside the United States. The practicality of this second mitigation requires that DoD have an accurate survey of such operations and locations and does not account for DoD operations outside fixed U.S. installations—for example, during a disaster assistance mission. Harmful interference may also be mitigated by a more interference tolerant design of Iridium terminals used by DoD. More interference tolerant designs may rely on coding gains, current Iridium high-bandwidth features may make this difficult and ultimately infeasible. As a practical matter, absent significantly more information on Iridium receiver design and Ligado transmitter characteristics near 1627.5 MHz accompanied by actual laboratory and field testing, it is impossible to predict the effectiveness of these technical mitigations. As a result, addressing this issue is most likely to require cooperative discussions among DoD, Iridium, and Ligado.

2.3.2 Outside the United States

The Ligado operation is licensed only in the United States, so there should be substantially no impact from the Ligado system to any DoD devices or operations outside the United States.

---

[10] 47 CFR § 25.253(g)(1).

[11] See "Petition for Reconsideration from Iridium Communications Inc., Aireon LLC, Flyht Aerospace Solutions Ltd., and Skytrac Systems Ltd." in the Ligado proceeding, document ID 10522231459169, received May 22, 2020, and posted May 26, 2020, p. ii, https://www.fcc.gov/ecfs/search/search-filings/filing/10522231459169, copy provided to the committee by Iridium on September 2, 2021.

[12] Letter submitted to the committee on February 22, 2022, which is available on request from the National Academies' Public Access Records Office.

[13] This paragraph was revised after this report's initial public release in September 2022 to clarify that certain assertions were made by stakeholders or participants in the Ligado proceeding and that those assertions were not being made by the study committee.

3

Additional Matters

The study committee's statement of task (see Appendix A) contains three primary tasks, which were addressed in the preceding chapters. The statement of task also provides that the committee may address "other related issues the study committee determines relevant." In the course of its work, the committee has concluded that there are several important issues surrounding the technical and administrative processes used in the long saga that has led to the authorities granted in Federal Communications Commission (FCC) Order 20-48. The following observations are offered with the hope that future proceedings might provide more streamlined and optimized approaches to balancing the desire to protect and enable incumbents while maximizing the economic and operational benefits provided by new entrants into a given spectrum region.

3.1 TOWARD A BETTER MEANS OF ASSESSING HARMFUL INTERFERENCE

It is important to understand why a large number of different communities reached diametrically opposed opinions on the likelihood of "harmful interference" from the Ligado license on Global Positioning System (GPS) and mobile satellite services (MSS) operations. The FCC defines harmful interference as "[i]nterference which endangers the functioning of a radionavigation service or of other safety services or seriously degrades, obstructs, or repeatedly interrupts a radiocommunication service operating in accordance with [the International Telecommunication Union] Radio Regulations." The committee believes that, because of a lack of a quantifiable definition of harmful interference, this has led to a 15-year process with respect to Ligado that has tied up spectrum, as well as untold resources of the participants. This issue is being played out to a lesser degree in the case of C-band licensing, and it is reasonable to expect that this issue will come up again in future with respect to GPS and other services.

Although there are different interpretations of test data, these differences arise in no small part from different assumptions about the design and performance of the receivers. Most aspects of the interference analysis have commonly accepted engineering principles: free-space loss, a standard National Telecommunications and Information Administration (NTIA) terrain propagation model, representation of antenna patterns, appropriate stand-off distance, and so on. However, many of the participants had very different, at times unfounded, assumptions about receivers. Some based their analysis on the worst-case device present in the field, some based on the *bulk* of the GPS receiver population, and others considered what *could* be the performance of a state-of-the-art receiver. There has been regulatory reluctance to interfere in the marketplace by having government standards for receiver performance (outside of safety communities, such as aviation), and receivers have not historically been part of the regulatory process. However, the committee believes that the FCC, working together with industry and government participants, can come up with measured levels for harmful interference in a given band of spectrum based on the intended usage in that band and based on near past and state-of-the-art receiver performance in the presence of significant adjacent-band power.

Policy makers have a necessary role in protecting existing uses of the spectrum, and not "fossilizing" industries by freezing spectrum usage and therefore precluding future uses. Unfortunately, "refarming," "repurposing," or "reallocating" spectrum (in this case, permitting a medium-power terrestrial adjunct) will have impacts for the existing users. As an example, this report identifies where this impact is likely for certain high-cost, high-accuracy civil GPS receivers that were designed prior to the application by Ligado. Regulators must be free to balance the impact on existing users with the benefits of future use. Existing users, however, should be provided some policy guarantees about the use of this equipment and service as they consider the investment. This policy should be broad, and not tied to specific transactions. There is a need for policy that balances existing users and future uses.

In the case of GPS, the committee believes that a sensible criterion for harmful interference could be developed that accounted for position error effects, acquisition and tracking challenges, and continuity of service. Such a criterion might be based on a maximum limit for degradation of C/N_0, and factor the possible effects of out-of-band emissions (OOBE) and adjacent-band signals in a designated frequency range *for a reasonably well designed receiver*. This analysis would then dictate an adjacent-band power mask that the FCC would guarantee going forward *for a given period of time*. Given this, it would then be possible for receiver manufacturers to guarantee acceptable receiver performance for cold-start acquisition, tracking, continuity of service, and positioning, navigation, and timing accuracy. This could be extended to other safety or communications services.

Another advantage of such an approach is that there would be no need to perform tests on large sets of receivers in order to evaluate whether harmful interference was occurring, either by a C/N_0 loss standard, a position accuracy degradation standard, or some suitable combination of the two approaches. An adjacent-band emitter would be guaranteed not to cause harmful interference in compliant GPS receivers as long as it complied with the maximum power limits for both its in-band and its OOBE.

If the FCC were to adopt future guarantees about adjacent-band power masks, the masks could be changed over time as receiver design technology improved. If the promise were, for example, for a 15-year period, then every 5 years a revised future mask guarantee could be published for the coming 15-year future interval over which no mask guarantee had yet been made. This would enable GPS receiver manufacturers to design, build, and sell devices to customers who would have a guaranteed period of service from their products.

Unlike much consumer and even industrial electronic equipment, certain GPS receivers are not throw-away devices that users will upgrade each time a new feature is added. High-end GPS receivers can be expensive and represent a major capital investment for some of their users—for example, surveyors, farmers, construction companies, and geodesists. These users cannot afford to buy these devices if they can expect only a 5-year service lifetime. The FCC has stated that it is "the responsibility of the receiver itself to incorporate a reasonable degree of adjacent-channel rejection in its design."[1] However, this is true only if receiver designers have a reasonable expectation of near-adjacent interference levels. Going forward, the FCC could forewarn GPS receiver manufacturers about what levels of adjacent-channel rejection will be needed in the future so that manufacturers can design, build, and sell suitable receivers.

Such a policy would also benefit new entrants to the use of adjacent bands. They would know exactly what was and was not permissible for at least a 10-year window. They could make proposals for changes to the next 5-year tranche of the allowed adjacent-channel power mask. GPS receiver manufacturers could then debate and negotiate about the next 5-year tranche, and the FCC could make a ruling based on the available technology going forward. Given the minimum 10-year horizon for the impact of any allowed changes, no party would suffer a bad shock.

The MSS band will likely continue to come under increasing pressure for expanded use. A proactive approach to the issues associated with how additional portions of the

---

[1] FCC Order 20-48, p. 34.

spectrum can transition from satellite to terrestrial use will help achieve outcomes that are fair to customers, current license holders, and other entities that have an interest in the band for non-MSS use.

3.2 WAYS TO MANAGE POTENTIAL FUTURE CONTROVERSIES

3.2.1 Receiver Standards

Consistent with the discussion in Section 3.1, the root cause of many of the current spectrum controversies arises from receiver designs that are predicated on different environments than would emerge after FCC rulemakings. This is not a simple "yes or no" problem, as there are significant issues with both the lack of standards and the imposition of standards. In many cases, the problem of receiver co-existence may be addressed through private arrangements among the parties, and this should not be precluded by regulatory solutions.

Many spectrum conflicts could be avoided if receivers were better designed and implemented. The natural inclination is that mandating higher levels of receiver performance would therefore resolve many of the spectrum conflicts that have arisen, and thereby would have value to industry and individual consumers. It is not the intent of this report to propose or oppose any specific action by the FCC. A reader of this report might be inclined to perceive that mandatory receiver standards might have precluded this controversy. Therefore, some other factors that should be considered include the following:

- The lifetime of many devices is shorter than the time in which major changes in spectrum are identified, considered and accepted, and then implemented. For example, smart watches, cell phones, and similar consumer equipment may go through many generations during the time it takes to consider even one spectrum usage change.
- The cost of high-performance analog filters is not subject to Moore's law. The weight, space, and cost of high-performance filters and the power requirements for high dynamic range receiver front ends are significant, and would likely preclude many of the technology products that have created so much social value for consumers and capability for industry.
- Many of the contentious spectrum actions have represented an extreme change in usage. For example, in the MSS and C-Band, the proposed use of the band went from 5 W transmitters, 40,000 km away, to nearby base stations with radiated powers thousands of times higher. While receiver

standards might have been valuable in these proceedings, implementing receiver standards across all users of the spectrum to deal with changes in utilization on the order of thousands of times more energy would place a heavy burden on all users, in order to simplify a small number of regulatory actions. One must thus distinguish between receiver standards that address operation in the current environment and those that might protect operation in the presence of some future uses that were vastly different in their impact.

One reason the FCC was faced with so much competing and conflicting input from the various communities in this proceeding is the lack of common receiver assumptions, which could provide a common point of departure for analytic efforts. Assumptions about receiver performance would be highly beneficial in focusing the discussion, without impacting the marketplace or equipment cost and performance. There is little point in having proceedings focus on analytic conclusions when there is no common basis for these analyses.

3.2.2 Ensuring Spectrum Succession

Technology, societal needs, and economics change. However, the spectrum management process treats each of its actions as stand-alone and immutable. It is essential that all spectrum decisions recognize that, at some point in time, they will be adjusted or changed completely. Spectrum processes must reflect that the environments will change, impacting both users of the immediate spectrum, and adjacent bands. Issues such as the varying equities inherent in the authorization of the Ligado system or in the controversies surrounding the implication of the 5G rollout on radio altimeters are not outliers and are not unique; they will become commonplace as new uses emerge and compete with older ones. A cohesive policy about rights of current users, the impact of equipment lifetime, business models, and all other considerations is essential, and should be established outside the pressures of any one spectrum decision. One potential means of systematizing this process is outlined in Section 3.1.

It is often U.S. government policy to promote technology succession (analog to digital television; 1G, 2G, 3G, and now 4G and 5G). But current actions supporting the repurposing of spectrum is at best ad hoc, and certainly does not operate on the same timeline as that which the technology is capable of changing. In the cellular world, spectrum has been repurposed between cellular generations by those holding the commercial licenses. Industry has accepted that it is disruptive to some users and that it obsoletes existing and even some high-cost equipment (e.g., cars with 3G cellular connectivity). These are reasonable decisions, weighing the capability of new uses against some disruption to the old ones. Yet, on the government side of spectrum decisions, there have been vigorous and effective defenses against changes, and a perceived obligation of any changed regime to not impact any of the existing uses. Although the study committee did recognize that at least one commercial piece of equipment might be negatively impacted in some deployments, the vendor has not manufactured equipment with this susceptibility for more than a decade. Yet, the fact that it was once sold (the committee could not determine how many, if any, were still in use) was sufficient for some opponents of the Ligado to justify opposition.

3.2.3 Administrative Process

In the course of its information gathering, the committee inquired of the FCC witnesses why, given the objections they raised regarding some positions provided in FCC filings, the FCC apparently did not consider introducing less rigorous protections than were asked for (yet more rigorous than are currently in the ruling). The FCC response was that they could not do so because no one had filed those specific requests. FCC regulatory decisions have both a policy component and a technical fact-finding one. The process appeared to the committee to be resolving questions of fact—for example, will Ligado interfere with GPS?—through administrative and/or procedural processes rather than a technical one. Although the committee is keenly aware of the constraints inherent in the Administrative Procedure Act,[2] it is the committee's opinion that selecting from specific filings may not be an appropriate technique to answer questions of fact. This proceeding might have had a more accepted outcome if the FCC was in a position to provide its own positions on factual questions. Parties have an interest in filing the most polarized positions that still might pass a "reasonableness" test. Limiting the FCC to picking between these positions is not amenable to the necessary good science for some of the questions in this proceeding.

3.2.4 Economic Mitigation

The FCC order established a process and responsibility for claims of interference to U.S. government equipment, and the mitigation of these issues through Ligado-funded replacements. Putting aside the question of whether there will be interference to GPS equipment, there are no clear policies for what rights equipment owners have with regard to future spectrum regulatory changes. What is the lifetime for any mitigation responsibility? What is the responsibility for receiver performance? How is any such responsibility addressed administratively? These are not issues that should be addressed on a case-by-case basis, but they should have some overarching policy support in advance of the resolution of specific issues.

---

[2] U.S. Congress, 1946, "Administrative Procedure Act (5 USC Subchapter II), §§ 551–559, Washington, DC.

The committee anticipates that spectrum repurposing will be increasingly necessary to support new technology and to support the continuing modernization of the U.S. economy, industry base, and other societal needs. Different FCC actions have had different impacts on incumbent uses. In some cases, such as the C-band auction, existing users were largely made whole through a variety of mechanisms, primarily based on auction proceeds as a responsibility of auction winners. In the Ligado proceeding, U.S. Department of Defense users were promised to be made whole, while there was little consideration for commercial equipment owners. Clearly, this lack of a "safety net" for existing users drove some of the opposition to the FCC actions.

Clearer understanding of the obligations of advocates for spectrum regulatory change, and of the rights of existing users, might go a long way to reduce much of the contention in similar proceedings in the future. Because much of U.S. spectrum policy is based on market principles of "highest and best use," it is important that the process provide a well-defined mechanism to identify and address the economic and other externalities of these uses.

3.2.5 Toward a More Collaborative Model of the Spectrum Process

The previous section discussed the inherent limitations of resolving spectrum issues through a process more akin to litigation than to engineering judgment. A similar issue arises from the separation of spectrum management responsibilities in the United States between federal and non-federal spectrum. This isolation of decision making may have been appropriate at some point in the past, when non-federal spectrum issues were dominated by broadcast, and federal spectrum issues were dominated by channel assignments for military operations. But today, there is little isolation between these domains. The federal government is a major user of commercial services and has a strong interest in the operation of these services.

A useful step to meeting the U.S. government objectives (as compared to the individual objectives of the FCC and the NTIA) would be to jointly study and test the impact of proposed regimes. Criteria would be agreed in advance, experiments agreed by all parties to be the relevant and inclusive cases. The government has very good engineers; they need to be given a chance to work out how to create the results that would enable policy makers to understand the options in meeting these two objectives simultaneously.

The committee believes that consideration of major changes in spectrum usage that impact both communities should have a higher level of coordination between the agencies, so that engineering and analysis is jointly directed. It is much easier to set criteria and determine the adequacy of proposed testing before the results are available. This interagency coordination would be above and beyond that provided in the legally mandated process, and it would work to create harmony in how U.S. national interests in spectrum policy are executed, even as disagreements are resolved.

It is important to note that many commercial spectrum rights issues have been resolved directly between the parties through some negotiated framework, which presumably balances the costs and risks to the participating parties. The FCC has greatly encouraged this process in the past, as regulators do not have to choose "winners and losers"; presumably, market forces resolve the issue. This did not happen during the previous LightSquared era, nor in the current Ligado proceeding, as agencies' processes defined the proceedings and those processes make different trades between benefits and risks. These trades are inherently negotiable decisions, and negotiation should be both accepted and encouraged, instead of endless reiteration of the same issues through successive cycles of argument.

4

Concluding Thoughts

The long history leading to the April 2020 Authorization of the Ligado Networks, LLC, low-power mobile satellite services (MSS) with an ancillary terrestrial component (ATC) stands as clear evidence of both the complicated technical, business, and operational challenges involved and of the complex administrative process that the Federal Communications Commission (FCC) has executed to reach its conclusions. The tasks assigned to the study committee are similarly complex, and the responses depend on the physics of the relevant electromagnetic emissions and reception, the use cases of that spectrum in a very busy bandwidth geography, and the implications of the long history of assessing and assigning spectrum rights for new entrants without generally assigning performance requirements on receivers.

In this context, the committee summarizes its responses to the task questions assigned as follows:

Task 1: [Consider] which of the two prevailing proposed approaches to evaluating harmful interference concerns—one based on a signal-to-noise interference protection criterion and the other based on a device-by-device measurement of the Global Positioning System (GPS) position error—most effectively mitigates risks of harmful interference with GPS services and U.S. Department of Defense (DoD) operations and activities.

> *Conclusion 1: Neither of the prevailing approaches to evaluating harmful interference concerns effectively mitigates the risk of harmful interference.*

Task 2: [Consider] the potential for harmful interference from the proposed Ligado network to mobile satellite services including GPS and other commercial or DoD services, including the potential to affect DoD operations, and activities.

> *Conclusion 2: Based on the results of tests conducted to inform the Ligado proceeding, most commercially produced general navigation, timing, cellular, or certified aviation GPS receivers will not experience significant harmful interference from Ligado emissions as authorized by the FCC. High-precision (HP) receivers are the most vulnerable receiver class, with the largest proportion of units tested that will experience significant harmful interference from Ligado operations as authorized by the FCC.*

> *Conclusion 3: It is within the state-of-the-practice of current technology to build a receiver that is robust to Ligado signals for any GPS application, and all GPS receiver manufacturers could field new designs that could co-exist with the authorized Ligado signals and achieve good performance even if their existing designs cannot.*

> *Conclusion 4: Iridium terminals will experience harmful interference on their downlink caused by Ligado user terminals operating in the UL1 band while those Iridium terminals are within a significant range of a Ligado emitter—up to 732 m.*

DoD has evaluated the impact of FCC Order 20-48 on department devices and missions and has asserted (see slides dated March 15, 2022[1]), "The terrestrial network authorized by FCC Order 20-48 will create unacceptable harmful interference for DoD missions. The mitigation techniques and other regulatory provision in FCC Order 20-48 are insufficient to protect national security missions."

Additional information on the test results and analysis as they related to DoD systems and missions is discussed in a classified annex to this report.

Task 3: [Consider] the feasibility, practicality, and effectiveness of the mitigation measures required in the FCC order with respect to DoD devices, operations, and activities.

> *Conclusion 5: Although the mitigation procedures proposed in the order may be effective, in many cases such mitigation may be impractical without the extensive dialog among the affected parties presumed in the Order. In some cases, mitigation may not be practical at operationally relevant time scales or at reasonable cost.*

---

[1] These slides, along with other materials provided to the committee, were placed in the project's public access file and are available on request from the National Academies' Public Access Records Office.

This report concludes with several additional observations related to the processes employed in the kinds of proceedings that led to FCC Order 20-48. Spectrum real estate is a living asset and approaches must allow not only for a degree of confidence that a deployed system will not be compromised by future, unforeseen entrants, *for a period of time*, but also must recognize that capabilities will evolve. Some form of more definitive receiver standards and establishment of set time periods where adherence to those receiver standards will ensure successful operation for a frequency band's incumbents and new entrants seem to be important tools in this regard.

Appendixes

A

Statement of Task

As requested in Section 1663 of the Fiscal Year 2021 National Defense Authorization Act, an ad hoc committee of the National Academies of Sciences, Engineering, and Medicine will provide "an independent technical review of the order and authorization adopted by the Federal Communications Commission on April 19, 2020 (FCC 20-48)," which authorized Ligado Networks, LLC, to operate a low-power terrestrial radio network adjacent to the Global Positioning System (GPS) frequency band.

The study will consider:

1. Which of the two prevailing proposed approaches to evaluating harmful interference concerns—one based on a signal-to-noise interference protection criterion and the other based on a device-by-device measurement of the GPS position error—most effectively mitigates risks of harmful interference with GPS services and U.S. Department of Defense (DoD) operations and activities.
2. The potential for harmful interference from the proposed Ligado network to mobile satellite services including GPS and other commercial or DoD services, including the potential to affect DoD operations, and activities.
3. The feasibility, practicality, and effectiveness of the mitigation measures proposed in the FCC order with respect to DoD devices, operations, and activities.

The committee's final report(s) will include the National Academies' findings and recommendations with respect to these issues as well as other related issues the study committee determines relevant.

The bulk of the technical analysis is expected to be performed based on public reports and open science and engineering literature and practice and result in an entirely unclassified report. This unclassified report is also expected to provide most or all of the analytical framework needed for assessing classified systems and capabilities.

Appropriately cleared members of the study committee will receive classified briefings and, if they determine it necessary, will prepare a classified annex.

B

Presentations to the Committee

SEPTEMBER 20, 2021

Fred Moorefield, Office of the Secretary of Defense, Chief Information Officer

OCTOBER 28, 2021

Doug Smith, Ligado
Valerie Green, Ligado

NOVEMBER 4, 2021

Charles Cooper, National Telecommunications and Information Administration (NTIA), U.S. Department of Commerce
Ed Drocella, NTIA
Scott Patrick, NTIA

NOVEMBER 18, 2021

Dennis Roberson, Roberson and Associates
Bill Alberth, Roberson and Associates
John Grosspietsch, Roberson and Associates

DECEMBER 9, 2021

Karen Van Dyke, Volpe Center, U.S. Department of Transportation (DOT)
Stephen Mackey, Volpe Center, U.S. DOT
Hadi Wassaf, Volpe Center, U.S. DOT
Christopher Hegarty, The MITRE Corporation
Karl Shallberg, Zeta Associates

DECEMBER 16, 2021

Adam Wunderlich, National Advanced Spectrum and Communications Test Network
Duncan McGillivray, National Advanced Spectrum and Communications Test Network
Dan Kuester, National Advanced Spectrum and Communications Test Network

JANUARY 6, 2022

Brad Parkinson, Stanford University

JANUARY 12, 2022

Scott Burgett, Garmin International
Stuart Riley, Trimble, Inc.

JANUARY 13, 2022

Scott Scheimreif, Executive Vice President, Government Programs, Iridium
Brandon Hinton, Engineering Consultant, Spectrum Analysis, LLC
Mark Settle, Senior Engineering Advisor, Wilkinson, Barker, Knauer, LLP
Ian Flood, Transfinite Systems Limited, London

JANUARY 20, 2022

Dana Goward, Resilient Navigation and Timing Foundation
Jordan Gerth, University of Wisconsin–Madison and Chair of AMS Committee on Radio Frequency Allocation
Renee Leduc, Narayan Strategy and Member of AMS Committee on Radio Frequency Allocations
Timothy Burch, National Society of Professional Surveyors (NSPS)
John "JB" Byrd, Federal Lobbyist, NSPS
Ed Hahn, Air Line Pilots Association
Bryan Lesko, Air Line Pilots Association
Andrew Roy, Aviation Spectrum Resources, Inc.
Sai Kalyanaraman, Collins Aerospace
John Shea, Helicopter Association International
Scott Bretthauer, National Agricultural Aviation Association

JANUARY 27, 2022

Ron Repasi, Office of Engineering and Technology (OET), Federal Communications Commission (FCC)
Paul Murray, OET, FCC
Steve Jones, OET Laboratory, FCC

C

Organizations and Individuals That Provided Written Input to the Committee

Aviation Spectrum Resources, Inc.
Brad Parkinson, Stanford University
CNH Industrial
Federal Communications Commission
Garmin International
GPS Innovation Alliance
Iridium Communications, Inc.
Ligado Networks
Michael Marcus, Northeastern University
Narayan Strategies
National Advanced Spectrum and Communications Test Network
National Society of Professional Surveyors
Office of Spectrum Management, National Telecommunications and Information Administration, U.S. Department of Commerce
Office of the Chief Information Officer, U.S. Department of Defense
Resilient Navigation and Timing Foundation
Roberson and Associates
Trimble, Inc.
UNAVCO
Volpe Center, U.S. Department of Transportation
WBK Law on Behalf of Garmin Ltd.

D

Committee Member Biographical Information

J. MICHAEL McQUADE, *Chair*, is a strategic advisor at Carnegie Mellon University, where he recently stepped down as the vice president for research. From 2006 to 2018, Dr. McQuade served as the senior vice president for science and technology at United Technologies Corporation, where he provided strategic oversight and guidance for research, engineering, and development activities that focused on a broad range of high-technology products and services for the global aerospace and building systems industries. Dr. McQuade has also held senior positions with technology development and business oversight at 3M, Imation, and Eastman Kodak. He served as the vice president of 3M's Medical Division and the president of Eastman Kodak's Health Imaging Business. Dr. McQuade has broad experience managing basic technology development and the conversion of early-stage research into business growth. He has served as a member of the President's Council of Advisors on Science and Technology, the Secretary of Energy Advisory Board, and the Defense Innovation Board, and is a member on the National Academies' National Science, Technology, and Security Roundtable and its Protecting Critical Technologies for National Security Consensus Study Report Committee. Dr. McQuade received a Ph.D. in physics from Carnegie Mellon University.

JENNIFER L. ALVAREZ is currently the chief executive officer and chair of the board for Aurora Insight, Inc. Alvarez co-founded Aurora Insight and started as the chief technical officer (CTO) in 2017. She led the development of the foundational technologies for sensing the radio frequency environment from land, air, and space, with a particular focus on emerging 5G technologies and the challenges associated with spectrum, including its general lack of availability and interference. Aurora Insight provided Ligado Networks,

Inc., with consulting services in 2018. Prior to co-founding Aurora Insight, Alvarez served in numerous roles at the Southwest Research Institute beginning in 1992. She performed extensive work on Global Positioning System (GPS) interference, including developing detection and mitigation techniques and systems for GPS jammers and spoofers. Alvarez led multi-disciplinary research efforts to develop innovative solutions in radio frequency signal acquisition and processing for cooperative, non-cooperative, and interfering signals. Her innovations included new radio frequency sensing technologies for cognitive radio applications and advanced methods for passively sensing and extracting information about signals and waveforms. Alvarez holds a B.S. in electrical engineering from The University of Texas at Austin and an M.S. in electrical engineering from The University of Texas at San Antonio, where her thesis focused on developing a communication technique that exploited certain characteristics of GPS signals.

KRISTINE M. LARSON is a professor emerita at the University of Colorado Boulder and a research associate at Central Washington University. Dr. Larson's research has focused on using GPS signals for geoscience research. In addition to using GPS to measure global and regional crustal deformation, she has developed new applications for GPS, including measuring seismic displacements, soil moisture, vegetation water content, precise time and frequency synchronization, snow depth, volcanic ash, tsunami waves, tides, and lake levels. Dr. Larson has served on several National Academies' committees, including the Committee on National Requirements for Precise Geodetic Infrastructure. She received the 2015 Huygens Medal from the European Geosciences Union. In 2020, Dr. Larson was elected to the National Academy of Sciences and received the Whitten Medal from the American Geophysical Union. She earned a B.A. in engineering sciences from Harvard University and a Ph.D. in Earth sciences (geophysics) from the Scripps Institution of Oceanography at the University of California, San Diego.

JOHN L. MANFERDELLI is currently the confidential computing incubation leader at VMware, where he leads security innovation projects in the office of the CTO. Before that, Dr. Manferdelli was a professor of the practice and the executive director of the Cyber Security and Privacy Institute at Northeastern University. Previous to that, he was the engineering director for production security development at Google. Before working at Google, Dr. Manferdelli was a senior principal engineer at Intel Corporation and the co-principal investigator (with David Wagner) for the Intel Science and Technology Center for Secure Computing at the University of California, Berkeley. Prior to working at Intel, he was a distinguished engineer at Microsoft and an affiliate faculty member in computer science at the University of Washington (UW). Preceding his work at Microsoft and UW, Dr. Manferdelli founded a natural language company that was acquired by

Microsoft and worked at Bell Labs, Livermore Labs, and TRW. He was a member of the Information Science and Technology Study Group at the Defense Advanced Research Projects Agency (DARPA) and is currently a member of the Defense Science Board and the National Academies' Forum on Cyber Resilience. Dr. Manferdelli's professional interests include cryptography and cryptographic mathematics, combinatorial mathematics, operating systems, computer security, and the Internet of Things. He is also a licensed radio amateur (AI6IT). Dr. Manferdelli holds a bachelor's degree in physics from Cooper Union for the Advancement of Science and Art and a Ph.D. in mathematics from the University of California, Berkeley.

PRESTON F. MARSHALL is an engineering director at Google, LLC. He is responsible for spectrum access technology, including a focus on the creation of a vibrant ecosystem of equipment, users, and standards in the newly shared 3.5 GHz Citizens Broadband Radio Service (CBRS) band. Dr. Marshall is chair of the Wireless Innovation Forum Spectrum Sharing Subcommittee, developing the technology base for 3.5 GHz spectrum sharing, and chair of the board of directors of the CBRS (now OnGo) Alliance, developing co-existence and neutral host technology for the 3.5 GHz band. He has a new book on this subject, *Three Tier Shared Spectrum, Shared Infrastructure, and a Path to 5G*, recently released by Cambridge University Press, as well as two prior works on cognitive radio. Dr. Marshall has been heavily involved in wireless technology and policy, holding positions such as deputy director of the Information Sciences Institute (ISI) at the University of Southern California (USC), director of the ISI Center for Computer Science and Technology, research professor at USC's Department of Electrical Engineering, contributor to the President's Council of Advisors on Science and Technology spectrum study that led to the creation of the CBRS band, and program manager at DARPA, directing multiple wireless and sensing technology programs. He earned a Ph.D. in electrical engineering from Trinity College, Dublin, and an M.S. and a B.S. in electrical engineering from Lehigh University.

MARK L. PSIAKI is the Kevin T. Crofton Faculty Chair in Aerospace and Ocean Engineering at Virginia Tech, where he has taught since 2016. Additionally, he is a professor emeritus of mechanical and aerospace engineering at Cornell University, where he taught for 30 years. Dr. Psiaki spent two sabbatical leaves as a Lady Davis Visiting Associate Professor with the Aerospace Engineering Faculty at the Technion in Haifa, Israel, and one sabbatical leave as a National Research Council senior research associate at the Space Vehicles Directorate of the Air Force Research Laboratory in Albuquerque, New Mexico. His contributions lie in the areas of estimation, filtering, data fusion, and signal detection with applications to GPS/global navigation satellite system navigation; alternative navigation methods; spacecraft attitude and orbit determination; and remote sensing of the upper atmosphere. Dr. Psiaki has authored or co-authored more than 80 refereed journal articles, 75 additional conference papers and trade magazine articles, 1 book chapter, and 10 U.S. patents. He is a fellow of both the Institute of Navigation (ION) and the American Institute of Aeronautics and Astronautics (AIAA). Dr. Psiaki has received the ION's Kepler, Tycho Brahe, and Burka awards, the Technion's Meir Hanin International Memorial Prize, and six best paper awards for AIAA conferences. He earned a B.A. in physics (1979), an M.A. in mechanical and aerospace engineering (1984), and a Ph.D. in mechanical and aerospace engineering (1987), all from Princeton University.

RICHARD L. (RICK) REASER, JR., has been a self-employed, independent consultant since January 2020. Reaser provides independent consulting services to U.S.-CREST on GPS and general space technology and markets. In spring 2021, he provided independent consulting services to Cerberus Operations and Advisory Company in the performance of technical and market due diligence to assess the viability of an aviation innovation. Reaser provided independent consulting services to the National Aeronautics and Space Administration (NASA) through the Aerospace Corporation with a systems engineering program assessment of the Artemis program in spring 2020. From 2006 to 2019, he led Raytheon Space and Airborne Systems' Spectrum Management and Electromagnetic Environmental Effects Department. Reaser was an Air Force officer from 1978 to 2006, when he retired as a Colonel. While in the Air Force, he served in the Air Force's GPS Joint Program Office for 12 years across three duty tours as a satellite engineer, satellite contract manager, chief engineer, and deputy system program director. As the U.S. Department of Defense (DoD) deputy director of spectrum management, Reaser was detailed by the Deputy Secretary of Defense to the White House and the U.S. Department of State as a technical advisor to the U.S. Ambassador to the World Radiocommunication Conference (WRC). In the late 1990s, he was selected as the U.S. spokesperson and leader of the interagency effort to prevent GPS spectrum encroachment. Reaser helped the United States and Europe obtain new international spectrum for GPS and Galileo at two WRCs (2000 and 2003). He negotiated the technical agreement between the European Union and the United States to share spectrum between the two systems in 2004. Reaser led the design efforts for three new GPS civil signals: L1C, L2C, and L5, as well the new military signal called M-Code. He was appointed by the Secretary of Commerce in January 2009 to the Commerce Spectrum Management Federal Advisory Committee as a Special Government Employee, where he served for a decade. In 2015, Reaser was selected by the National Academy of Sciences to serve on a congressionally directed committee that provided scientific, technical, and management recommendations regarding the U.S. Department of Commerce's telecom labs.

JEFFREY H. REED is the Willis G. Worcester Professor in the Bradley Department of Electrical and Computer Engineering at Virginia Tech. Dr. Reed currently serves as the founding director of Wireless@Virginia Tech, one of the largest and most comprehensive university wireless research groups in the United States, which he founded in 2006. In 2010, Dr. Reed founded the Ted and Karyn Hume Center for National Security and Technology and served as its interim director. From 2019 to 2020, he served as the interim director of the Commonwealth Cyber Initiative, and he is the current CTO. His area of expertise is in software radios, smart antennas, wireless networks, and communications signal processing. Dr. Reed received his Ph.D., M.S., and B.S., all in electrical and computer engineering, from the University of California, Davis (respectively 1987, 1980, and 1979). He has participated in various National Academies' activities, including a U.S. Government Accountability Office expert meeting on broadband deployment in 2016, and in reviews of engineering and research activities at the Institute for Telecommunications Sciences and the Communications Technology Laboratory of the U.S. Department of Commerce in 2015. Additionally, Dr. Reed has served on the technical advisory boards for approximately six companies and as an informal advisor on national policy regarding wireless issues.

NAMBIRAJAN SESHADRI is currently a professor of electrical and computer engineering at the University of California, San Diego, a distinguished visiting professor at the Indian Institute of Technology-Madras, India, and serves as an advisor to a number of start-ups. Dr. Seshadri is a fellow of the Institute of Electrical and Electronics Engineers (IEEE), a member of the National Academy of Engineering (2012), and a recipient of the IEEE Alexander Graham Bell Medal (2018). He received his M.S. and Ph.D. from Rensselaer Polytechnic Institute in 1984 and 1986, respectively, and his B.E. from the University of Madras (Regional Engineering College, Tiruchirappalli, India) in 1982. Dr. Seshadri began his career at AT&T (Bell Labs from 1986 to 1995 and Shannon Labs from 1996 to 1999), where he conducted research in various aspects of mobile radio systems, culminating in the invention of space-time codes for which he and his co-authors won the best paper award from IEEE Transactions on Information Theory. He joined Broadcom in 1999, and was the CTO of the Mobile and Wireless Division from 2000 to 2016, as well as general manager of the mobile group from 2011 to 2015.

J. SCOTT STADLER is the head of the Communication Systems mission at the Massachusetts Institute of Technology Lincoln Laboratory. He directs a portfolio of programs spanning architecture definition, technology development, system prototypes, and testbeds that are advancing the capabilities of the nation's communication networks. The division focuses on military satellite communications, free-space laser communications, ground- and air-based tactical radios, spectrum operations, and the development of quantum technologies for communications. Dr. Stadler has been involved in the design, development, and operation of a number of NASA and DoD satellite systems both at Lincoln Laboratory and in industry. This work included the early design and prototype of an architecture for supporting packetized network services via satellite. He has also led research efforts focused on the seamless integration of wireless and terrestrial packet data networks. Dr. Stadler has served in a variety of technical management positions at Lincoln Laboratory and also served as the chief engineer for the AF SMC/MCX through the Intergovernmental Personnel Act and as a member of the Air Force Scientific Advisory Board. He holds a B.S. from Worcester Polytechnic Institute, an M.S. from University of Southern California, and a Ph.D. from the University of Pennsylvania, all in electrical engineering.

STEPHEN J. STAFFORD is the chief scientist of the GPS and GNSS Group at the Johns Hopkins University Applied Physics Laboratory (JHUAPL). Stafford is a member of JHUAPL's principal professional staff, holding a B.S. and an M.S. in electrical engineering from the University of Maryland, College Park, and the University of California, Berkeley, respectively. He has more than 15 years of experience in position, navigation, and timing sensor fusion as well as radionavigation, including the development of several GPS receivers for high-accuracy and weak signal applications. Stafford has written several publications related to the field of navigation warfare.

E

Disclosure of Unavoidable Conflicts of Interest

The conflict of interest policy of the National Academies of Sciences, Engineering, and Medicine (http://www.nationalacademies.org/coi) prohibits the appointment of an individual to a committee authoring a Consensus Study Report if the individual has a conflict of interest that is relevant to the task to be performed. An exception to this prohibition is permitted if the National Academies determine that the conflict is unavoidable and the conflict is publicly disclosed. A determination of a conflict of interest for an individual is not an assessment of that individual's actual behavior or character or ability to act objectively despite the conflicting interest.

Mark L. Psiaki has a conflict of interest in relation to his service on the Committee to Review FCC Order 20-48 Authorizing Operation of a Terrestrial Radio Network Near the GPS Frequency Bands because he consults for and receives research support from NAL Research Corporation, which designs and manufactures communications systems that use the satellite network owned and operated by Iridium Communications, Inc. The National Academies have concluded that for this committee to accomplish the tasks for which it was established, its membership must include at least one expert who possesses substantial current expertise in Global Positioning System (GPS)/global navigation satellite systems signals processing, radionavigation, and mobile satellite services. As described in his biographical summary, Dr. Psiaki has extensive practical experience in mobile satellite services (MSS) and radionavigation and GPS and global navigation satellite system signal processing, including his current advisory work on MSS device solutions at NAL Research Corporation. The National Academies have determined that the experience and expertise of Dr. Psiaki is needed for the committee to accomplish the task for which it has been established. The National Academies could not find another available individual with the equivalent experience and expertise who does not have a conflict of interest. Therefore, the National Academies have concluded that the conflict is unavoidable. The National Academies believe that Dr. Psiaki can serve effectively as a member of the committee, and the committee can produce an objective report, taking into account the composition of the committee, the work to be performed, and the procedures to be followed in completing the study.

Nambirajan Seshadri has a conflict of interest in relation to his service on the Committee to Review FCC Order 20-48 Authorizing Operation of a Terrestrial Radio Network Near the GPS Frequency Bands because of his ownership of stock in Verizon Communications. The National Academies have concluded that for this committee to accomplish the tasks for which it was established, its membership must include at least one expert who possesses substantial recent experience in the engineering of cellular phones and portable GPS devices. As described in his biographical summary, Dr. Seshadri has extensive recent experience in research and engineering leadership positions at several communications technology firms, where he led practical implementation of both cellular and GPS services in chipsets and handheld devices. The National Academies have determined that the experience and expertise of Dr. Seshadri is needed for the committee to accomplish the task for which it has been established. The National Academies could not find another available individual with the equivalent experience and expertise who does not have a conflict of interest. Therefore, the National Academies have concluded that the conflict is unavoidable. The National Academies believe that Dr. Seshadri can serve effectively as a member of the committee, and the committee can produce an objective report, taking into account the composition of the committee, the work to be performed, and the procedures to be followed in completing the study.